METALORGANIC CHEMIC

DEPOSITION FOR THE HETEI

HOT ELECTRON DI

METALORGANIC CHEMICAL VAPOR DEPOSITION FOR THE HETEROSTRUCTURE HOT ELECTRON DIODE

by

Mark A. Emanuel

University of Illinois
Coordinated Science Laboratory
Urbana, Illinois

NOYES DATA CORPORATION
Park Ridge, New Jersey, U.S.A.

Library of Congress Catalog Card Number: 88-32954
ISBN: 0-8155-1195-7
Printed in the United States

Published in the United States of America by
Noyes Data Corporation
Mill Road, Park Ridge, New Jersey 07656

10 9 8 7 6 5 4 3 2 1

Library of Congress Cataloging-in-Publication Data

Emanuel, Mark A. (Mark Andrew)
Metalorganic chemical vapor deposition for the heterostructure hot electron diode.

Bibliography: p.
Includes index.
1. Diodes, Semiconductor. 2. Gallium arsenide semiconductors. 3. Vapor-plating. I. Title.
TK7871.86.E44 1989 621.3815'2 88-32954
ISBN 0-8155-1195-7

In Memory of My Father

Foreword

Metalorganic chemical vapor deposition (MOCVD) is an epitaxial crystal growth technique capable of producing high-quality compound semiconductors in thick or thin layers with abrupt junctions, excellent areal uniformity, and precisely controlled thickness, doping and composition. In this book the desired characteristics of an MOCVD system are described, and design criteria necessary for their implementation are identified. Special emphasis is placed on defensive design strategies intended to limit the extent of system perturbation due to various component failure modes and normal maintenance procedures. The design of reactor computer control software is also considered, and algorithms for the growth of layers graded in both doping and composition are presented.

Specifically, theory and experimental data are presented for the heterostructure hot electron diode (H^2ED), a new two-terminal electronic device that exhibits S-shaped negative differential resistance (NDR) in the dc I–V characteristic.

The information in the book is from *Metalorganic Chemical Vapor Deposition and Its Application to the Growth of the Heterostructure Hot Electron Diode,* by Mark A. Emanuel of the University of Illinois Coordinated Science Laboratory, for the Joint Services Electronics Program, the National Science Foundation, and the Department of Defense, March 1988.

The table of contents is organized in such a way as to serve as a subject index and provides easy access to the information contained in the book.

Advanced composition and production methods developed by Noyes Data Corporation are employed to bring this durably bound book to you in a minimum of time. Special techniques are used to close the gap between "manuscript" and "completed book." In order to keep the price of the book to a reasonable level, it has been partially reproduced by photo-offset directly from the original report and the cost saving passed on to the reader. Due to this method of publishing, certain portions of the book may be less legible than desired.

Acknowledgments

I thank first my advisor, Professor James Coleman, for his technical guidance, support and encouragement during this work. I also thank Professors Karl Hess, Thomas DeTemple, Oscar Gaddy, James Kolodzey, Gregory Stillman, John Tucker and Joseph Lyding for helpful discussions.

The help of my fellow students in performing this work was greatly appreciated. Ted "Dorian" Higman, whose device characterization talents and thorough knowledge of the ecclesiastical modes of the Western scale never failed to amaze me, was instrumental in this work. Mike Givens was always there to beat the equipment into submission. Linda Miller forced me to understand MOCVD well enough to explain it to her, and her help with the crystal growth was much appreciated. Greg Costrini gave me my first instruction in MOCVD. Jim Baillargeon unselfishly offered his points of view in a variety of areas. Pam York colored the lab in interesting ways. Chuck Zmudzinski always made sense. Gloria Fernandez assisted me in reading papers in the Spanish-language technical journals. Rob Bryan and I had many interesting discussions just out of curiosity. Mike Favaro was never unkempt. Kevin Bernink relieved me of concern that the faith might not be kept. Luke Mawst assisted me in ways too numerous to detail. Chris Mondragen and Diana Carroll were helpful in keeping me current in the literature. Sylvia Patterson was of assistance in her own special way.

Special thanks go to Rob Thorne for many inconclusive discussions of the meaning of it all.

Many others have helped make my time here something very special in one way or another. Jerome Hubacek, Steve Skala, Greg Lyons, Roger Brockenbrough, Jack Higman, Doug Arnold, Isik Kizilyalli, Dan Bailey, Joy Laskar, Steve Manion, Virginia Robbins and Michael Haase all deserve a special mention.

For technical support, I thank Keith Kuehl, Al Wilson, Yuri Moroz and the gentlemen at the EE Machine Shop. I also thank Ottie Johnson for her patience.

Finally, I thank my family for continually urging me to finish school and get on with my life. A mother's advice is always best.

This work has been supported by the Joint Services Electronics Program, the National Science Foundation Materials Research Laboratory, the National Science Foundation Engineering Research Center for Compound Semiconductor Microelectronics, and the Department of Defense University Research Instrumentation Program.

NOTICE

The material in this book was prepared as an account of research sponsored by the Joint Services Electronics Program, the National Science Foundation, and the Department of Defense. On this basis the Publisher assumes no responsibility nor liability for errors or any consequences arising from the use of the information contained herein. Mention of trade names or commercial products does not constitute endorsement or recommendation for use by these Agencies or the Publisher.

Final determination of the suitability of any information or procedure for use contemplated by any user, and the manner of that use, is the sole responsibility of the user. The reader is warned that caution must always be exercised when dealing with semiconductor raw materials and processes, and expert advice should be sought at all times before implementation is considered.

Contents and Subject Index

1. Introduction

Metalorganic chemical vapor deposition (MOCVD)[1-3] is an epitaxial crystal growth technique that is capable of producing high quality compound crystalline material in thick or thin layers with abrupt junctions, precisely controlled thickness, doping and composition, and excellent areal uniformity. Materials systems to which MOCVD has successfully been applied in terms of both bulk material and device structures include III-V and II-VI compound semiconductors and IV-VI lead salts. Structures that have been successfully produced by MOCVD include electron real-space transfer devices,[4] high electron mobility transistors,[5,6] strained-layer superlattices[7] and double heterostructure,[8] complementary self-aligned[9] and quantum well lasers.[10]

Chapter 1 of this thesis is a brief overview of the MOCVD process with specific attention given to the aluminum gallium arsenide material system used in this work. Chapter 2 is a discussion of various aspects of MOCVD reactor design and maintenance with emphasis on reactor reliability considerations. Chapter 3 is a description of the reactor control algorithms developed for the growth of thin layers and layers graded in both composition and doping. Finally, Chapter 4 is a presentation of results and discussion of investigations of the heterostructure hot-electron diode,[11] a new device exhibiting S-shaped negative differential resistance.

1.1 Introduction to MOCVD

MOCVD is a vapor phase epitaxial growth technique in which at least one of the crystal component precursors is a metal alkyl. Precursors, also called reactants or sources, are transported in a carrier gas to the vicinity of a hot substrate where a thermally driven growth reaction takes place. A simplified diagram showing the essential components of an MOCVD reactor is given in Fig. 1. Such reactors are

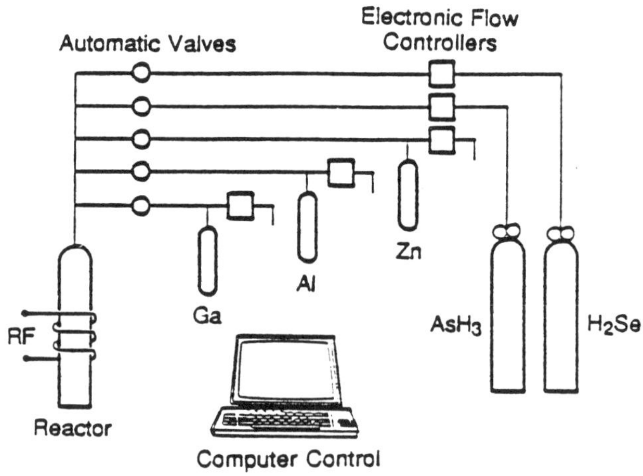

Figure 1. Simplified diagram of an MOCVD reactor system equipped to grow doped $Al_xGa_{1-x}As$ structures.

constructed of stainless steel tubing, automatic bellows valves and electronic mass flow controllers. Gaseous precursors are contained in high pressure gas cylinders, and metal alkyl sources are contained in special cylinders called bubblers that allow sources vapors to be transported by a carrier gas, typically ultrapure hydrogen. Computer control is often used to ensure precise and reproducible control over all process parameters.

For the case of a binary III-V crystal grown with metal alkyl sources, the net growth reaction is given by

$$(R_1)_pM_{III} + (R_2)_qM_V \rightarrow M_{III}M_V + R_3H \qquad (1.1)$$

where R_1 and R_2 are organic radicals, M_{III} is a column III metal, M_V is a column V metal, $M_{III}M_V$ is the desired crystal compound and R_3 is an organic by-product formed from R_1 and R_2. In general, there is an excess of the column V element in the gas phase in order to prevent thermal damage to the substrate and growing layer by the outdiffusion of the column V element. The reaction is thus limited by the availability of the column III metal. Also, the column V constituent is usually supplied by a metal hydride rather than a metal alkyl precursor. For the specific case of the growth of gallium arsenide (GaAs) by the reaction of trimethylgallium (TMGa) and arsine (AsH_3), the net growth reaction then becomes

$$(CH_3)_3Ga + AsH_3 \rightarrow GaAs + 3CH_4 \qquad (1.2)$$

where methane (CH_4) is a stable organic by-product. The growth of compound crystals with more than two column III constituents is accomplished by adjusting the gas phase stoichiometry such that the relative proportions of the column III species are the same as those desired in the crystal, and a similar but independent adjustment is necessary in the case of crystals containing multiple column V constituents. This simple rule can be complicated by such factors as differences in the temperature dependence of decomposition among reactants and parasitic reactions such as those involving indium alkyls and hydrides.[12] For the well-behaved aluminum gallium arsenide ($Al_xGa_{1-x}As$) system, the net growth reaction becomes

$$x(CH_3)_3Al + (1-x)(CH_3)_3Ga + AsH_3 \rightarrow Al_xGa_{1-x}As + 3CH_4 \qquad (1.3)$$

which is Eq. (1.2) with trimethylaluminum (TMAl) added.

The growth rate of an epitaxial layer is a linear function of the column III precursor input rate, and the achievable range is determined by precursor vapor pressure and associated gas flow control hardware. Growth rates can typically be varied on the order of 1 to 100 Å/s, with the lower growth rates used for the precise growth of thin layers. Layers with thicknesses ranging from a few monolayers to a few tens of microns can be reproducibly grown.

1.2 Column III precursors

The most commonly used precursors for gallium and aluminum are trimethylgallium and trimethylaluminum, respectively. They are commercially available in high-purity form, and have vapor pressures high enough to allow the controlled and reproducible transport of their vapors by a carrier gas. In addition, they participate in no parasitic reactions with other commonly used precursors in the AlGaAs material system, and they do not interact with the materials used in reactor construction in such a way as to produce turn-on or turn-off transients. There are, however, some considerations that have caused alternative precursors to be investigated. Methyl radicals produced during the decomposition of methyl metal alkyls may decompose further at typical growth temperatures to yield carbon that may be incorporated into the growing layer. By this mechanism, carbon incorporation into AlGaAs increases with increasing growth temperature.[13] Also, lower growth temperatures may be necessary to prevent significant thermal disordering or dopant redistribution during growth in some device structures,[14] or to minimize some thermally driven parasitic reactions.[12] These considerations have led to the investigation of ethyl metal alkyls, which allow growth at lower temperatures and yield a more stable organic radical upon decomposition, as alternative precursors.[15] Triethylgallium (TEGa) and triethylaluminum (TEAl) have been shown to be viable alternatives to the corresponding methylated compounds in many applications. A major impediment to its widespread use, particularly for TEAl, has been their extremely low vapor pressure which necessitates elevated source temperatures or extremely large gas flow rates through the source bubblers to achieve reasonable growth rates. Its use also decreases the incorporation of carbon into the epitaxial layer from decomposed organic radicals in both GaAs and AlGaAs. The problem of carbon incorporation is particularly severe for aluminum containing compounds owing to the strength of the aluminum-carbon chemical bond, and this has prompted a search for aluminum precursors that mitigate this problem. Dimethyl aluminum hydride has been used as an aluminum precursor for the growth of AlGaAs having very low carbon contamination.[16]

Indium precursors have been the topic of much investigation because of the propensity of indium alkyls to form addition compounds with column V hydrides,[12] resulting in an effective removal of a significant fraction of the indium precursor from the reactant stream. The extent of this parasitic reaction has been observed to vary from run to run, resulting in irreproducibility in the growth of critically lattice-matched layers, such as InGaAs on an InP substrate.[12] Both trimethylindium and triethylindium have been used, but successful application of either precursor is critically dependent upon reactor design and growth condition parameters. An alternative approach is the use of preformed adducts as sources.[17] It is believed that since the adduct is already formed, no further parasitic reactions will occur in the gas stream, resulting in a known and reproducible delivery rate of the precursor. Some success has been achieved with this approach, although it is limited by its inability to independently vary the relative proportions of the adduct components.

1.3 Column V precursors

Arsine and phosphine are typically used as the arsenic and phosphorous precursors, respectively. They are available in high pressure gas cylinders and are usually used either pure or as 10% of a hydride-hydrogen mixture. Electrically active contaminants are not detected in high-purity arsine, although there is indirect evidence of the presence of zinc compounds.[18] Water and oxygen are frequently present, and these impurities have extremely deleterious effects on the quality of aluminum-containing alloys.[19] Tanks are usually specified to have less than one part per million each, and further steps may be taken to reduce the concentration of these impurities in the gas before it is used, such as the use of a molecular sieve or indium-aluminum-gallium eutectic bubbler.[20]

The extremely high toxicity of arsine and phosphine, coupled with the fact that they are stored in high pressure gas cylinders, has spurred the search for less toxic, more easily handled metal alkyl column V sources. Experimental investigations of arsenic precursors such as trimethylarsenic,[21] diethylarsine[22] and tertiarybutylarsine[23] have yielded GaAs layers with surface morphology or impurity concen-trations inferior to those achievable by use of arsine. It is unclear whether these results are due to impurities contained in the sources or to unfavorable growth reaction chemistry.

1.4 Doping

Doping is accomplished through the introduction of a suitable dopant precursor into the growth chamber during the growth of the epilayer. Dopant incorporation is usually a linear function of the precursor input rate up to the solid solubility limit. Several elements have been used as donors in III-V semiconductors by MOCVD, including column VI elements sulfur,[24] selenium,[25] tellurium[26] and column IV elements silicon[26] and tin.[28] Selenium and silicon are the most commonly used. Selenium is useful when heavy n-type doping is required, as electron concentrations greater than 10^{19} cm^{-3} can be achieved through its use, but the commonly used precursor for selenium, hydrogen selenide (H_2Se), strongly adsorbs onto internal reactor surfaces. Desorption from these surfaces after the hydrogen selenide flow is terminated results in a gradual decrease of the precursor concentration in the gas stream. Such behavior is referred to as a memory effect, and can result in doping transients that are unacceptable in some devices.[29] In addition, the use of hydrogen selenide results in a strong growth temperature dependence of dopant incorporation which can affect layer doping uniformity and reproducibility.[25] Silicon is a useful dopant owing to its low diffusion coefficient in III-Vs and its low vapor pressure, which permits it to be used in planar doping applications.[30] Although amphoteric, it does not incorporate significantly as an acceptor at concentrations less than the maximum achievable electrically active doping level, approximately 5×10^{18} cm^{-3} in GaAs.[27] In addition, the commonly used precursor for silicon, silane (SiH_4), is used in vast quantities in the silicon semiconductor industry and thus is readily available in high-purity form. However, as with hydrogen selenide, silane exhibits a very strong growth temperature dependence for decomposition and thus is susceptible to similar problems of uniformity and reproducibility, as well as extremely poor doping efficiency at low growth temperatures.[27] As an alternative to silane for silicon doping, disilane (Si_2H_6) has been used.[27] It has a much lower thermal decomposition energy than silane and displays no temperature dependence of doping over the growth temperature range of 600-850° C.

P-type dopants for III-V semiconductors by MOCVD come from columns IIa and IIb of the periodic table. Zinc,[31] magnesium[32,33] and beryllium[34] have been used successfully as p-type dopants. Zinc is commonly used in the form of diethylzinc ($(C_2H_5)_2Zn$), which is a well-behaved precursor having a conveniently high room temperature vapor pressure and no doping memory effects. Zinc doping can be used to give hole concentrations approaching 10^{20} cm^{-3} in GaAs, which makes it extremely useful for the growth of contact layers for nonalloyed contacts. Zinc, however, has an extremely high diffusion coefficient in GaAs.[32] It is thus susceptible to dopant redistribution during growth and subsequent high-temperature device processing. Magnesium, in the form of either bis(cyclopentadienyl) or bis(methylcyclopentadienyl) magnesium, is often employed in situations where a dopant with a low diffusion coefficient is necessary, as its diffusion coefficient is five orders of magnitude less than that of zinc. The maximum hole concentration achievable with magnesium is approximately $2x10^{19}$ cm^{-3}, which may be an unacceptable limitation in some device structures. Perhaps a more important consideration is the fact that doping with the magnesium precursors mentioned above results in very long turn-on and turn-off transients. Beryllium, in the form of diethylberyllium ($(C_2H_5)_2Be$), possesses the better characteristics of both zinc and magnesium, having a lower diffusion coefficient than magnesium, no significant doping transient and a high maximum achievable hole concentration. Unfortunately, beryllium compounds, particularly oxides, are extremely toxic and this has discouraged its widespread application in MOCVD.

1.5 Growth chambers, gas dynamics and uniformity

The actual growth of an epitaxial layer is accomplished by transporting the desired precursors in a carrier gas, usually ultra-pure hydrogen, into a growth chamber containing a heated susceptor. The precursors decompose in the vicinity of a substrate on the susceptor, and growth takes place. The uniformity of the grown layer is a complex function of the gas dynamics near the substrate, the carrier gas velocity, and, for the usual arsenic-rich growth conditions, the column III precursor input rate.[35]

There are two basic reactor configurations. The first, the horizontal reactor, is shown in Fig. 2 (a). It is a horizontal cylinder typically constructed of high-purity quartz. The precursor stream enters one end and flows toward a horizontal susceptor. The downstream edge of the susceptor is raised slightly to promote more uniform deposition by compensating for reactant depletion over the susceptor in the direction of gas flow. The gas exits the chamber through the other end. The second basic configuration, the vertical reactor shown in Fig. 2 (b), is a cylinder with a circular cross section into which the carrier gas and reactants enter through the top and flow downward toward a disk-shaped susceptor that is perpendicular to the incoming gas flow. The gas flows towards the hot substrate, the growth reaction takes place and the gas flows out the bottom of the chamber through the exhaust line. The susceptor is sometimes rotated to promote uniform deposition. In both configurations various types of susceptor heating systems have been used, including radio frequency induction, heat lamps and electrical resistance. For both the horizontal and vertical configurations, two main gas flow considerations must be taken into account. These are the overall gas flow patterns in the chamber and the gas flow near the substrate. For ease of visualization these will be described within the context of the vertical reactor.

The overall gas flow pattern in the chamber is dominated by the interaction of the gas with the heated susceptor, assuming that the susceptor rotation rate is low. Gas entering the top of the reaction chamber flows down toward the hot susceptor. As it nears the susceptor it is heated and expands, and the resulting buoyant force tends to counter its downward flow. If the initial downward flow is not large enough, then this buoyant force can dominate, causing a reversal of flow direction near the susceptor and setting up recirculation cells as shown in Fig. 3. This is an extremely undesirable situation, as the gas flow in the chamber is then nonuniform, resulting in nonuniform

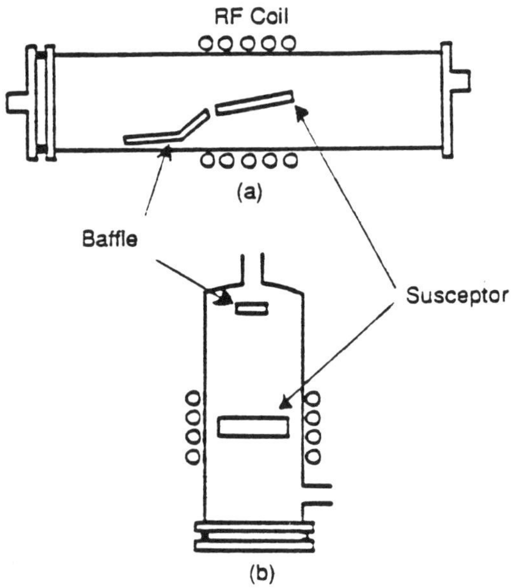

Figure 2. The (a) horizontal and (b) vertical reaction chamber configurations.

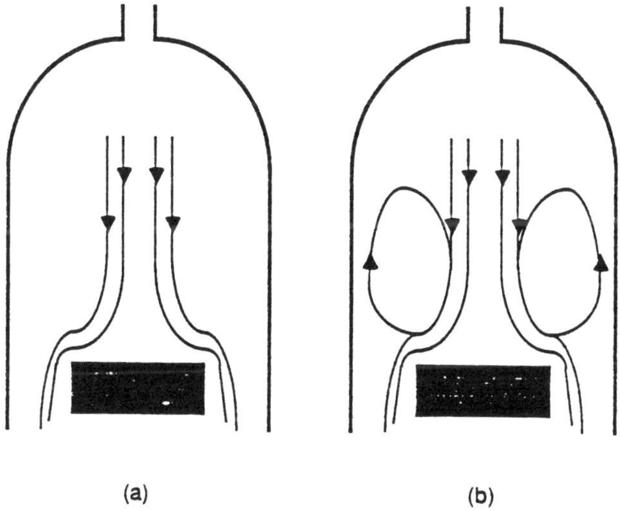

Figure 3. Flow patterns in a vertical reactor. The desired well-behaved flow is shown in (a). Recirculation cells are shown in (b). (After reference 36)

layer growth. Also, recirculation cells can cause graded junctions by effectively storing a portion of the gas mixture for a particular layer and slowly releasing it after the input gas mixture has been changed for the growth of a different layer. The formation of these cells is a function of the carrier gas species and flow rate, the chamber configuration and susceptor temperature. In the case of a vertical reactor with high-speed susceptor rotation, the pumping effect of the rotating susceptor can dominate the buoyant forces and prevent the formation of the cells.[36] The gas flow pattern is difficult to observe directly, but qualitative studies have been performed using titanium dioxide smoke to reveal flow features.[36] Computer modeling of flow patterns has shown good agreement with experimental data on layer uniformity for some chamber geometries and growth conditions.[37]

As the gas flows over the hot susceptor a stagnant, or boundary, layer is formed which acts as a barrier through which reactants must diffuse in order to reach the substrate. This is illustrated in Fig. 4. This boundary layer plays a crucial role in the determination of the uniformity of the grown layer. Ideally, carrier gas flow rate and growth temperature are adjusted such that the boundary layer is of uniform thickness over the substrate. If the reactants above the boundary layer are uniformly distributed, as they will be in a thoroughly mixed gas stream, then the growth will be uniform. Variations in the boundary layer thickness will result in corresponding variations in reactant diffusion rate through the layer and thus nonuniform growth. When the boundary layer is adequately developed, the growth is said to be in the diffusion limited regime. This is the desired case as excellent uniformity can be achieved. When the boundary layer is not developed adequately, the growth is said to be in the mass-transport limited regime. In this case, growth occurs as fast as reactants are carried to the substrate. Good uniformity is difficult to achieve in this regime. Conditions for uniform growth must be determined empirically for a given reactor configuration. A systematic study of the problem has shown that with proper growth parameters layer

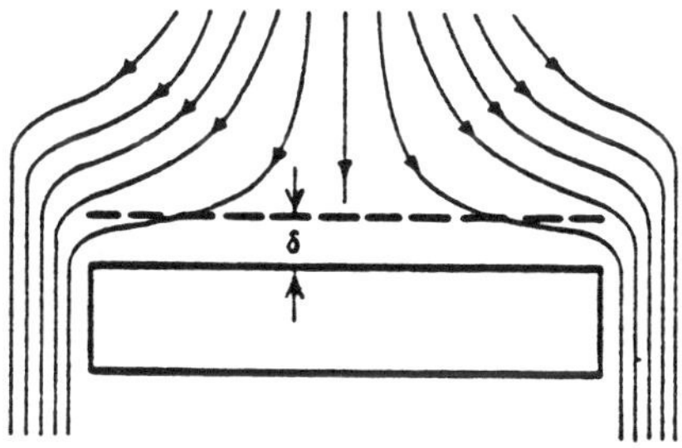

Figure 4. Gas flow in the vicinity of the susceptor in a vertical reaction chamber. The boundary layer thickness is indicated as δ.

thickness uniformities of better than ±1.5% over a 4 centimeter diameter substrate are achievable in a vertical reactor similar to that described above.[35]

1.6 Crystal quality considerations

The purity of the GaAs grown by MOCVD tends to be limited by impurities contained in the precursors rather than any intrinsic limitations in the process itself. It has been shown that the purity of nominally undoped GaAs grown by MOCVD can be improved by further purification of trimethylgallium received from vendors.[38] Major impurities have been shown to be germanium, silicon, tin, sulfur, tellurium, carbon and zinc.[39] The highest 77K mobility reported for MOCVD grown GaAs is 190,000 $cm^2/v\text{-}s$, which compares favorably to the maximum values obtained by other techniques.[39] As the demand for higher purity precursors increases and the metal alkyl market becomes more competitive, it is expected that vendors will become more responsive to the needs of those using MOCVD technology with a corresponding improvement in product purity.

Excellent surface morphology is easily achieved in layers grown by MOCVD provided that growth parameters, such as temperature and V-III ratio, are properly adjusted. Defect density in the grown layer, as determined by etch pit measurements, is limited by the defect density of the substrate. The effect of substrate defects on device performance can be reduced by use of graded or superlattice buffer layers.[40]

1.7 Heterojunctions

Atomically abrupt, high-quality heterojunctions are essential for a variety of device structures. Studies have shown MOCVD to be capable of producing abrupt, dislocation free interfaces in several material systems, including the lattice matched GaAs-AlAs system[41] and the strained layer GaAs-GaAsP system.[7] This is possible because under proper growth conditions MOCVD growth is dominated by two-dimensional layer by layer growth rather than three-dimensional island growth.[7] In addition, the growth kinetics of MOCVD are such that there are no etching reactions competing with the growth process. These facts ensure a smooth growing surface. Achieving abrupt interfaces then becomes a problem of ensuring no mixing of reactant gases between adjacent layers. This can be done by interrupting the growth after each layer is completed and purging the growth chamber of reactants before the growth of the next layer is started. In Fig. 5 is shown a high resolution transmission electron micrograph of a GaAs/AlAs superlattice with 17 Å layer thicknesses grown with a 15-second growth interruption between layers. The interfaces are smooth and abrupt to one monolayer.

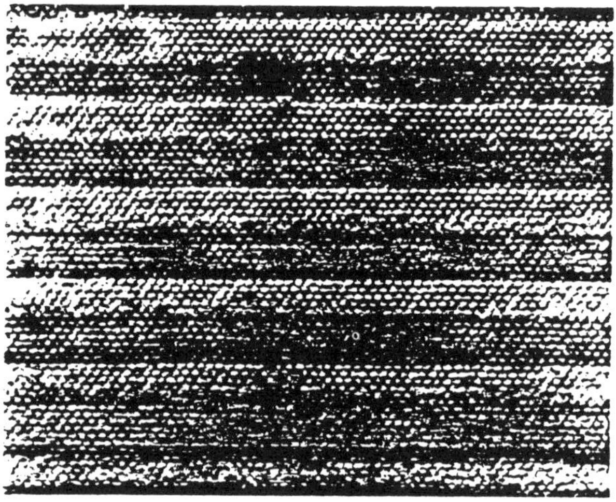

Figure 5. High resolution transmission electron micrograph of a GaAs/AlAs superlattice having a layer thickness of 17 Å. (After reference 42)

2. MOCVD Reactor Design Considerations

The growth of high quality crystals by MOCVD requires, in addition to an understanding of the growth process itself, a well-engineered reactor. This implies more than well-designed simply from a functional point of view. A well-engineered reactor produces material of good quality and continues to do so in spite of having been subjected to such adversities as component failures, source changes and normal maintenance procedures. Potential trouble spots must be identified and defensive design strategies implemented either to eliminate them or minimize their impact.

The design of an MOCVD reactor from several perspectives is described in this chapter. First, the basic requirements that the reactor must meet will be considered. Then the materials of construction, components and construction techniques necessary will be discussed. Next, a simple GaAs reactor design will be described, and the design will then be expanded to enable the growth of multilayered AlGaAs doped structures. Then, this design will be modified for maintenance considerations and then modified still further to minimize system damage in the event of component failure. Finally, human interface considerations will be discussed.

2.1 Reactor requirements

An MOCVD reactor must be capable of producing cystals with low background impurity level, good surface morphology, abrupt junctions and uniform thickness, composition and doping. The satisfaction of each of these requirements is dependent to some extent upon the reactor design and maintenance. Each will now be discussed within the context of its reactor dependence.

2.1.1 Background impurity level

Unintentional impurities in MOCVD grown material have several different origins: the reactant sources, the MOCVD growth process, the substrate and the reactor. In an optimized reactor system, source contamination has been shown to be a major source of background impurities,[38] and this problem can be addressed only through careful packaging of source materials. The MOCVD growth process itself can cause carbon from decomposed organic radicals to incorporate into the growing crystal. Since this impurity cannot be eliminated from the growth environment, this source of unintentional impurities can be addressed only through changes in growth conditions,[13] substrate orientation[42] or precursor species that cause changes in the growth kinetics which make the incorporation of carbon less favorable. The substrate itself has been shown to be a potential source of impurities,[43] and proper treatment of the substrate prior to growth has been shown to greatly mitigate this problem. Finally, the reactor itself, because of poor design and construction, can be a major source of impurities.

Proper materials and techniques for reactor construction are critical to the minimization of background impurities. Much of this aspect of MOCVD reactors is a carryover from silicon epitaxial reactors based on the pyrolysis of silane, and a summary will be presented here. The material of choice is 316L stainless steel because of its extremely low reactivity and ease of passivation. Materials such as brass and aluminum are not acceptable because of their relatively high reactivities and their potential for introducing contaminants into the gas stream. Seals and connections should be metal-inert gas (MIG) welded or metal gasket seals such as Cajon VCR. Solder and braze connections are unacceptable in high-purity systems. Wherever possible, soft seals should be made of inert plastic such as Teflon or Kel-F. Rubber seals should be avoided because they emit impurities and are difficult to clean. Where rubber seals are unavoidable, as in commercially available mass flow controllers, Viton is the material of choice. Proper cleaning of all materials prior to welding and assembly is essential. All parts should be degreased in organic solvents and then given a passivating treatment where appropriate. Cleaned stainless steel should be treated in dilute nitric acid to form a passivating chromium oxide layer.[44] Cleaned materials should be stored away from airborne contaminants such as oil vapor and human breath. Likewise, reactor assembly should be performed in a clean environment. Extreme care should be taken to ensure against the introduction of any contaminants during this process.

2.1.2 Surface morphology

Surface morphology problems in MOCVD have the following origins: improper substrate preparation, inadequate pumping and purging of the growth chamber after substrate loading, unfavorable growth kinetics, and particles originating in the reactor. Substrate preparation problems, while requiring some detective work, are usually easily resolved, and in any case cannot be addressed through reactor design. Unfavorable growth kinetics can often be corrected by altering growth conditions. But again, reactor design is not a significant factor. Particles originating in the reactor, however, are a problem that can be addressed through proper reactor design and maintenance.

Assuming that the reactor components have been properly prepared prior to assembly, the major source of surface morphology disturbing particles is the reaction of water and oxygen with metal alkyls. This reaction produces a nonvolatile film on internal reactor plumbing surfaces that acts as a source of particles. Water and oxygen can gain access to the reactor through contaminated gaseous reactant sources, leaks to atmosphere, hydrogen purifier failure, and maintenance and operation errors. In addition, particles are generated by the pyrolysis of unspent reactants downstream of the growing crystal. The growth chamber must be designed such that no recirculation of the downstream gas occurs in such a way as to deposit these particles on the growing crystal surface.

2.1.3 Uniformity

Areal uniformity of layer thickness, composition and doping is primarily dependent upon growth conditions, the uniformity of the susceptor temperature profile, and the design of the reaction chamber. Because of the hydrodynamic complexity of the system, both the chamber design and growth condition determination are typically done empirically. Excellent growth temperature uniformity can be achieved with a number of different susceptor heating systems, with radio frequency induction widely used owing to its freedom from electrical feed-throughs into the growth environment. The design of the reactant delivery system plays an important role in the uniformity problem in that thorough mixing of the reactant gases must take place before they reach the growth zone. This requirement can cause problems when using precursors that necessitate minimum contact time with one another in order to avoid parasitic reactions prior to their reaching the growth zone, and thus result in an inadequately mixed gas stream.

2.1.4 Abrupt junctions

The growth of multilayered structures with abrupt hetero- and doping junctions is primarily a reactor design problem. Recirculation cells in the growth chamber are a potential cause of imperfect junctions. The reactant delivery system can also cause nonabrupt junctions through design deficiencies that result in memory effects, reactant pulse interdiffusion and staggered reactant arrival times. This is discussed in Section 2.3.2.

2.2 Gallium arsenide reactor design

A schematic of a gallium arsenide reactor is shown in Fig. 6. Although very simple, this reactor is functionally adequate for the task of growing nominally undoped gallium arsenide. However, it possesses several inadequacies which make its long-term trouble-free operation unlikely. These will be discussed later. Nonetheless, it does contain all essential components and major design concepts and thus is useful for descriptive purposes. The reactor of Fig. 6 is sectioned into several subsystems: the carrier gas supply, the metal alkyl supply, the hydride supply, the reactant mixing manifold, the growth chamber, the exhaust scrubber and the laboratory safety system. Each will now be described.

2.2.1 Carrier gas supply subsystem

The function of the carrier gas is to transport the reactants to the substrate. Hydrogen, helium and nitrogen have been used as carrier gases in MOCVD. Hydrogen is by far the most commonly used because it is easily and economically purified, but there are situations where an alternate gas is desirable, such as in fundamental studies of MOCVD[45] and in the growth of some indium-containing compounds where nitrogen is used to reduce adduct-forming reactions by decreasing the thickness of the thermal boundary layer above the substrate and the overall gas temperature in the growth chamber.[46]

Hydrogen is the carrier gas used in the example system. Extremely clean gas is needed. It can be supplied in individual tanks of expensive, high-purity (99.9999%) gas, but the high rate of hydrogen usage makes this economically impractical. Perhaps more important is that each tank change exposes part of the plumbing to air, resulting in some degree of system contamination. Such problems are avoided by use of a commercially available hydrogen purifier which is a hot palladium thimble through which hydrogen, and only hydrogen, diffuses. Inexpensive, welding-grade hydrogen

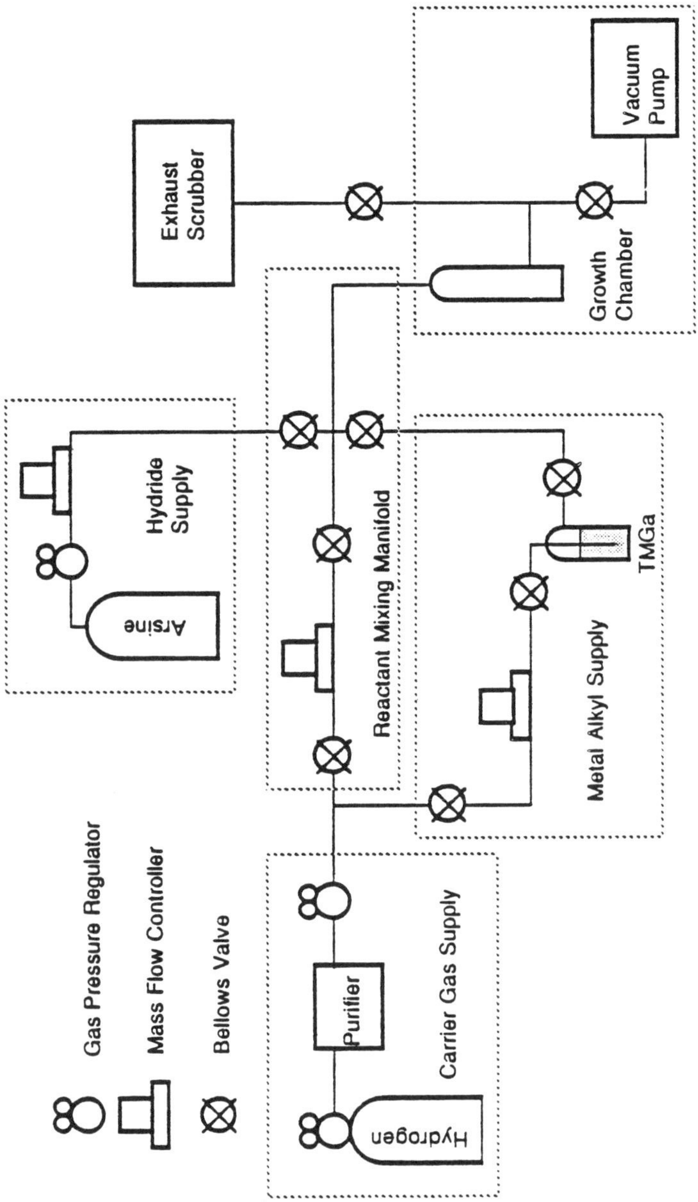

Figure 6. Major subsystems of a simple MOCVD reactor.

(99.9%) is passed through a regulator to give a delivery pressure suitable for the purifier, typically one- to two-hundred psig. Hydrogen diffuses through the palladium element and exits through the high purity port, and nonhydrogen components of the feed gas exit through a waste port. Halogens, hydrides, sulfur compounds and unsaturated hydrocarbons must not be present in the feed gas as these are damaging to the palladium.[47] Because the purifier delivery pressure will be a function of the hydrogen delivery rate, a high-purity regulator is used to deliver a constant hydrogen pressure to the reactor from the purifier, which is critical to the stable operation of electronic mass flow controllers.

2.2.2 Metal alkyl supply subsystem

The gallium source for the reactor is trimethylgallium. Like most other metal alkyls, it is available from vendors in a special container called a bubbler, which is a stainless steel cylinder with a valved input into which a metered flow of carrier gas is introduced, and a valved output through which the carrier gas and metal alkyl vapors exit. The input carrier gas is often directed to the bottom of the cylinder by a dip tube from where it bubbles upward through the contents. This facilitates the saturation of the unfilled volume at the top of the bubbler with metal alkyl vapor. Since the mass transport rate of the metal alkyl is a function of its vapor pressure as well as the carrier gas flow rate, the bubbler is usually immersed in a constant temperature bath to hold the source vapor pressure constant.

The metal alkyl supply subsystem consists of a connection to the hydrogen supply, an electronic mass flow controller to meter the hydrogen through the bubbler, the bubbler itself, valves to isolate the bubbler from the reactor during source changes and a shutoff valve for controlling reactant delivery to the manifold.

2.2.3 Hydride supply subsystem

Arsine (AsH_3), either in pure form or mixed with hydrogen, is the arsenic precursor used on the example system. Arsine is supplied by vendors in a high-pressure gas cylinder. It passes through a pressure regulator to bring the delivery pressure down to a value suitable for mass flow controller operation, typically 20 psi. The final component in this subsystem is a shutoff valve.

2.2.4 Reactant manifold subsystem

The reactant manifold, also called a mixing manifold, is the section of plumbing into which all reactants are introduced for transport to the reaction chamber. Because the reactant flows make up only a small fraction of the total gas flow necessary to establish an appropriate boundary layer above the substrate, there is typically a large carrier gas flow through the manifold to make up the difference. For this simple reactor, the mixing manifold is simply a cross into which the arsine, TMGa and the carrier gas are introduced. The fourth port, the output, is directed to the growth chamber. Such simplicity is possible only because with this reactor hetero- and doping junctions are not possible, and thus the abruptness of such junctions is not a concern. A reactant mixing manifold design capable of producing abrupt junctions will be discussed in Section 2.3.2.

2.2.5 Growth chamber subsystem

The growth chamber is the vessel in which the actual epitaxial growth takes place. It consists of an inlet for carrier gas and reactants, a susceptor upon which the the substrate is placed, an exhaust port, a heating and temperature measurement system for the susceptor and a vacuum pump for evacuating the reaction chamber for substrate loading and unloading. The two general classes of growth chambers, vertical and horizontal, were discussed in Chapter 1. A vertical reaction chamber is used in the example system. Reaction chambers are usually constructed of high-purity quartz, and the joint between the chamber and manifold line is usually sealed with a Viton O-ring. Because quartz cannot be formed to precise dimensional tolerances as metal can, the section of manifold tubing that attaches to the input connection of the chamber is frequently a flexible metal tube that allows for chamber-to-chamber variations in dimensions. An additional part of this subsystem is a vacuum pump for evacuating the chamber after substrate loading.

2.2.6 Exhaust scrubber subsystem

Neutralization of the reactor exhaust is of critical importance from a health and safety standpoint. The exhaust of the gallium arsenide reactor described here will consist mainly of hydrogen, arsine and arsenic dust, with small amounts of methane and gallium compounds. Arsine is an immediate threat to life, arsenic and its oxides are poisons that accumulate in the body over time, and hydrogen is explosive. These exhaust components must be disposed of in such a way as to endanger neither laboratory personnel nor the environment. A typical exhaust scrubbing system consists of a mechanical filter to capture particulates, followed by a treatment of the gas stream to remove arsine. This can be accomplished by bubbling the exhaust through a potassium permanganate solution, but a more common and economical approach has been the use of activated charcoal filters.[48] The remaining gas, which at this point would consist mostly of clean hydrogen, is frequently diluted with large amounts of air or nitrogen and exhausted through a tall stack. Another technique is the use incineration to reduce the toxic components of the exhaust to less toxic, more easily disposed of solids.

2.2.7 Laboratory safety subsystem

The use of extremely toxic gases in MOCVD dictates that every aspect of reactor design, operation and maintenance be carefully scrutinized for the possibility of accidental release of these gases into the laboratory environment. However, in the event of a toxic gas release, a safety system must be in place to prevent injury both to laboratory personnel and unsuspecting bystanders. Every possible leakage point for toxic gases must be enclosed in an exhausted cabinet kept at negative pressure relative to the laboratory. In addition, the laboratory itself must be sealed and kept at negative pressure relative to the rest of the building. The laboratory and exhaust system must also be thoroughly equipped with toxic gas monitors to provide immediate warning of any leaks. Toxic sources should be equipped with remotely operated valves that can be closed when trouble is indicated. There should be interlock switches on reactor cabinets to prevent reactor operation with open doors. Such a system naturally lends itself to computer control, but to avoid its failure in the event of an interruption in electrical service all safety related systems should be on a noninterruptable power system. Some elaborate safety systems have been described in the literature.[49,50]

2.3 Expansion of simple design for doped AlGaAs structures

Expansion of the simple gallium arsenide reactor just described for the growth of doped aluminum gallium arsenide structures obviously involves the addition of aluminum and n-type and p-type dopant precursors. However, the possibility of the growth of abrupt doping junctions, heterostructures and quantum wells necessitates a more sophisticated approach to plumbing design than was necessary for the simple reactor.

Successful growth of abrupt junctions is simply a matter of abruptly and controllably changing the reactant mixture stoichiometry immediately above the growing crystal. There are, however, several design issues affecting this goal that must be addressed.

2.3.1 The run/vent configuration

Reactant precursors must be switched in and out of the mixing manifold quickly. In addition, there must be no transients, such as pressure surges or delays, on turn-on or turn-off. Failure to meet these requirements can result in graded or otherwise imperfect junctions, compositional or doping transients near interfaces and inability to control layer thicknesses. The metal alkyl supply system shown in Fig. 6 is inadequate for the task at hand for a number of these reasons. The problem is that the flow in the supply line is not initiated until the alkyl is needed in the mixing manifold. A brief examination of the system reveals that no matter which valve or combination of valves is opened to allow alkyl into the mixing manifold, there will be a flow surge. This is due to the mass flow controller having pressurized the upstream line in an attempt to maintain its commanded flow while the valve was closed, or a delay before the reactants reach the manifold, or both. Note that a flow of 10 standard cubic centimeters per minute (sccm) at constant pressure in one-quarter inch outside diameter tubing gives a gas velocity of approximately one inch per second. In addition, the reactant vapor would be stagnant in the line prior to its injection into the manifold, and this could result in the accumulation of contaminants from the internal plumbing surfaces.

These problems are avoided by use of the run/vent configuration[29,51] shown in Fig. 7. In this system, long before the reactant is needed in the manifold, an equilibrium flow is established in the line and directed to a vent. When the reactant is needed in the manifold, the vent valve is closed and the reactant-to-manifold, or run, valve is opened simultaneously, thus directing the equilibrated reactant flow to the manifold. When the reactant is no longer needed in the manifold, the run and vent valves are again toggled simultaneously and the reactant is directed to vent. In order to ensure

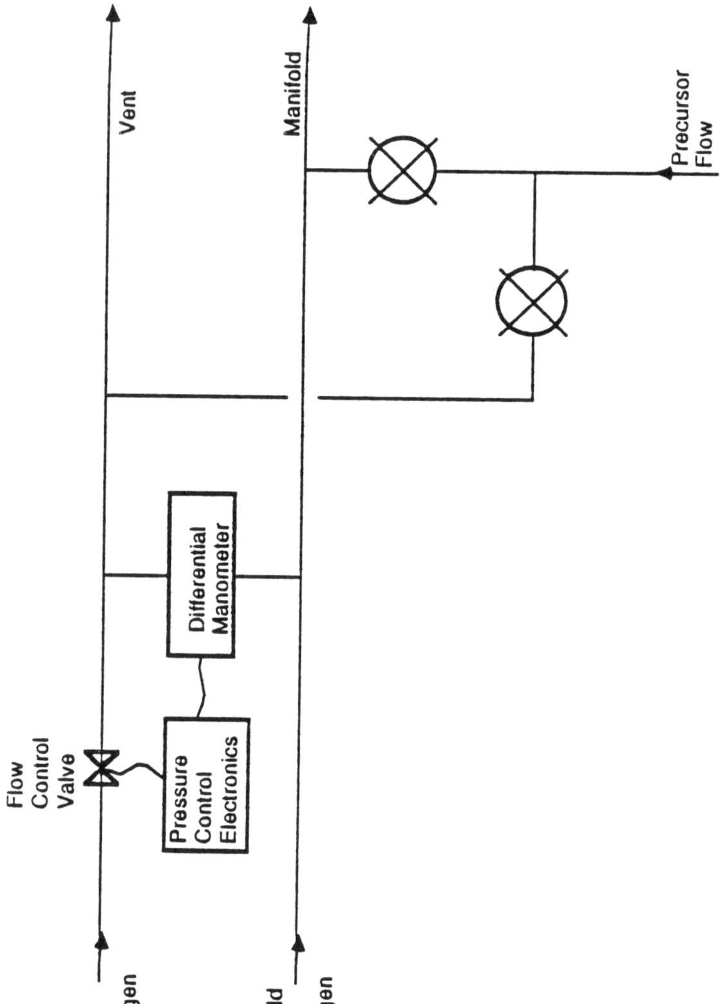

Figure 7. A schematic diagram of the run/vent configuration with automatic pressure balancing between the manifold and vent lines.

that the equilibrium flow is maintained during the switch, the pressures in the vent and manifold lines are often actively balanced by use of an automatic control system.[29,52] A pressure difference between these lines can have adverse effects on thickness control for thin layers and heterojunction quality. For example, assume that the pressure in the vent line is greater than that in the manifold. The pressure in a reactant line will equilibrate at the pressure of the line to which it is directed. A reactant line thus equilibrated to manifold pressure will be at a lower pressure than the vent, and when it is switched to the higher pressure vent the contents of the vent flow into the reactant line until the pressure is again equilibrated. Besides the obvious contamination problem with this situation, it may take several seconds for the gas pulled in from the vent to be purged. If the reactant line is switched back to manifold before the unwanted gas is cleared from the line, this unwanted gas will be sent to the manifold instead of the desired reactant. In addition, because the reactant line is at a higher pressure than the manifold, there will be a surge of gas from the reactant line, resulting in transients in doping or composition. Such effects have been observed for a pressure difference of as little as 3 Torr.[53]

2.3.2 Reactant mixing manifold

The design of the reactant mixing manifold is of critical importance for the growth of precision junctions. In the gallium arsenide reactor just described, the mixing manifold could be very simple as the problem of switching multiple reactants did not exist. With the multiple reactant system, however, new problems must be addressed.

The first of these problems is dead space,[54] which is unpurged or stagnant volume in the mixing manifold where significant quantities of reactant gases can accumulate. Dead space will fill with the reactant gases for the layer being grown, and when the reactant mixture is changed for the growth of the next layer the reactant mixture from the previous layer that has accumulated in the dead space will mix with the intended reactant mixture, causing a graded layer transition. An example of dead space in a manifold is shown in Fig. 8. This figure shows valves which control TMGa and TMAl delivery to the manifold. In Fig. 8 (a) the TMGa valve is open for the growth of GaAs. When the TMGa valve is closed and the TMAl valve is opened so that AlAs may be grown, the volume between the TMGa valve seat and the manifold line is dead space. As illustrated in Fig. 8 (b), this dead space will be filled with TMGa vapor at the same concentration that was being injected into the manifold. As an estimate of how significant this might be, consider that a typical GaAs growth rate is 6 Å/s for a TMGa flow of 10 sccm into the manifold, giving a figure of 0.083 scc per 3 Å monolayer. Now consider that 1/4 inch outside diameter (o.d.) tubing has approximately a 0.2 cm^2 internal cross section, and thus one inch of this tubing has a volume of 0.5 cm^3. This is enough to grow 6 monolayers of GaAs. This will very likely lead to a graded junction, although the exact profile of the grading will depend upon the rate at which the vapor in the dead space enters the manifold gas stream. An analysis of this situation by van Opdorp and Leys[54] indicates that the trapped TMGa will diffuse out of the dead space approximately exponentially in time, and that after 15 seconds more than 95% will have exited.

A less severe case of dead space is illustrated in Fig. 9. Consider a situation in which only the TMGa valve is open for the growth of GaAs. The entire manifold downstream of that valve will be filled with TMGa vapor, including the dead space associated with the TMAl and silane valve connection. In this case, however, the TMGa vapor trapped in the dead space has been diluted by the carrier gas. For a carrier gas flow of 10,000 sccm and TMGa flow of 10 sccm, the resulting dilution factor of 1000 renders this dead space much less significant than that described above for undiluted source vapor. Nonetheless, even this could be significant if dopant is trapped in dead space prior to the growth of a thin, high-purity GaAs layer.

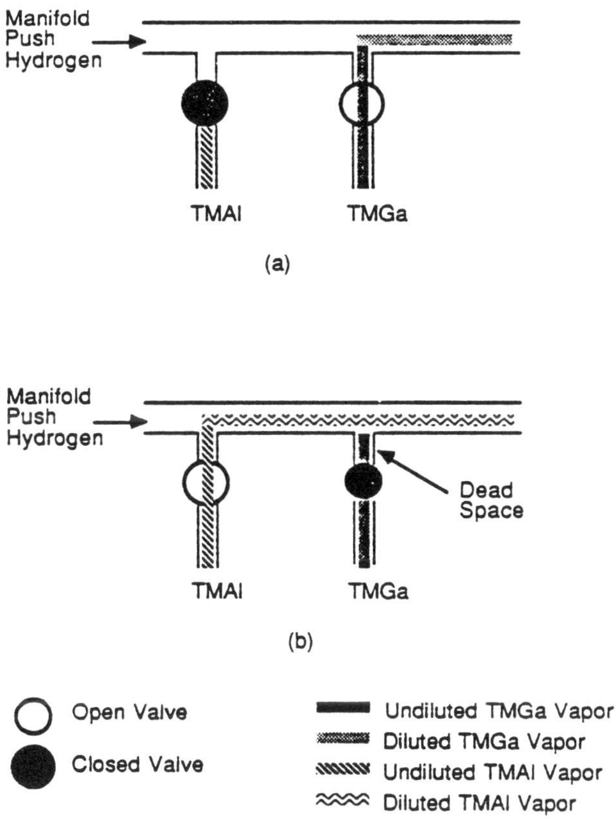

Figure 8. Illustration of dead space in a manifold. The transition from GaAs growth in (a) to AlAs in (b) will not be abrupt owing to the undiluted TMGa vapor trapped in the dead volume as indicated.

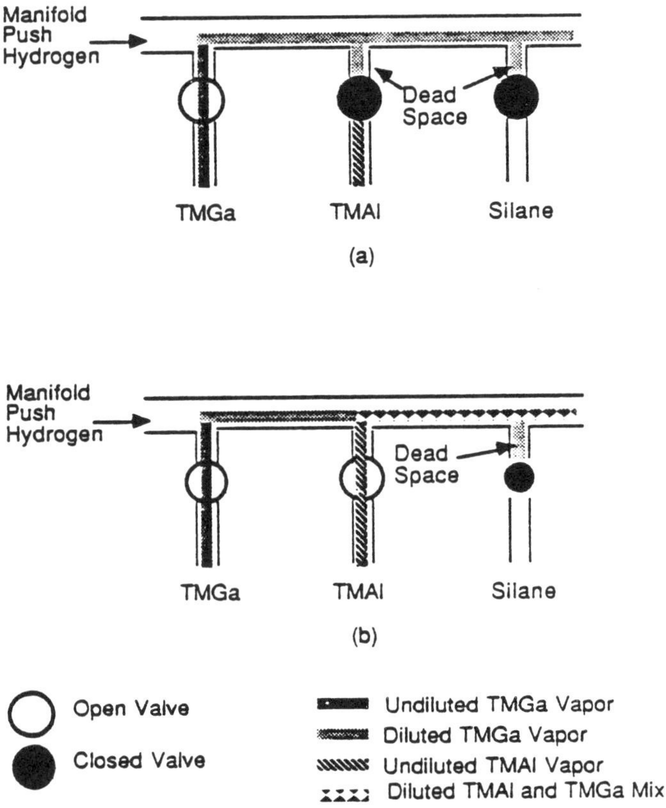

Figure 9. Illustration of dead space in a manifold. The indicated dead space on the transition from the growth of GaAs in (a) to AlGaAs in (b) is not as critical as that shown in Fig. 8 owing to the dilution of the trapped vapor by the manifold push hydrogen.

Dead space due to valve connections to a manifold is typically avoided by use of special three-port valves called manifold valves in which two ports are an unobstructed path through the valve body. The gas to be injected into the manifold enters through the third port, and its access to the manifold is controlled by a valve seal. Such valves essentially eliminate manifold-reactant valve connection dead space. A manifold valve is shown in Fig. 10.

Another manifold consideration is the transit time to the growth chamber for each reactant. Ideally, all reactants arrive simultaneously so as to avoid a situation such as silane arriving before TMGa at the start of an n-type GaAs layer. In this case, Si would deposit on the crystal surface until the TMGa arrived and growth started.[30] The resulting dopant sheet could adversely affect the electrical and structural qualities of the interface. Similar scenarios can be imagined for various combinations of precursors and arrival times. The use of the type of manifold pictured in Fig. 8, known as a linear manifold, results in staggered reactant arrival time. This problem can be avoided by use of a radial manifold in which all reactants are injected equidistantly from the chamber input port. Radial manifolds tend to be rather complex and bulky, while linear manifolds are easily assembled from commercially available parts, are more compact and are more easily modified. The timing problems inherent in the linear manifold can be greatly reduced by use of an extremely high carrier gas velocity. The difference in arrival time between two precursors is equal to the linear separation of their injection points in the manifold divided by the carrier gas velocity. Thus, for a separation of 2 inches and a gas velocity of 100 inches per second (10,000 sccm in one-quarter-inch o.d. tubing) there is a difference of arrival time of only 0.02 second, which is two orders of magnitude less than the time required to grow a single monolayer at typical growth rates. Such a small stagger would probably have negligible effect on an interface.

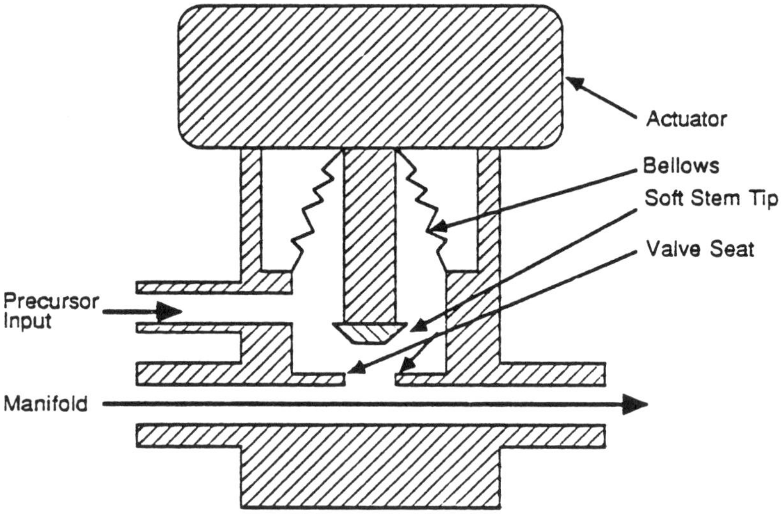

Figure 10. Cross section of Nupro Company manifold valve. Several such valves can be connected in series to construct a complete manifold.

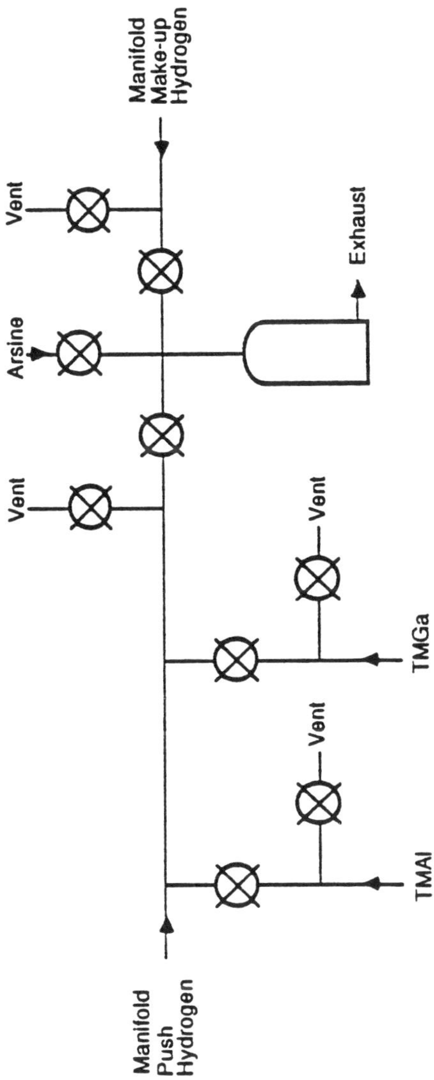

Figure 11. A schematic diagram of the "double run/vent" manifold configuration.

The problems of dead space and staggered precursor arrival times can be virtually eliminated by use of what can be called the double run/vent manifold shown in Fig. 11. In this configuration, the desired column III precursors are switched into the manifold carrier gas stream via the run/vent switching described above. However, rather than being sent to the chamber they are diverted to the vent, and an identical carrier gas flow containing no column III reactants is switched into the chamber to maintain an equilibrium flow. During this time the precursors in the manifold equilibrate, precursors trapped in dead volume diffuse into the manifold stream and are discarded, and the leading edge of the reactant stream, where precursors are staggered, is eliminated. After a suitable time, usually several seconds, the manifold is switched into the chamber and the next layer growth starts. The hydride flow to the chamber is never interrupted in order to maintain the column V element overpressure necessary to prevent substrate damage. The use of such an equilibrating pause between layers has the advantage of helping to ensure extremely abrupt interfaces. A disadvantage of this technique is that during the pause impurities in the growth chamber gas stream can accumulate on the surface of the crystal and be trapped at the interface when the next layer is grown.[30]

Another consideration is the amount of time that the precursors are in contact with one another prior to reaching the growth zone. Certain precursors react with one another to form nonvolatile addition compounds, thus effectively removing them from the growth process. Also, a long residence time in the manifold can allow interdiffusion of gas mixtures for consecutive layers, which can result in graded interfaces. Thus, to minimize their contact time, they should be mixed as closely as possible to the growth zone and the carrier gas velocity should be high. The latter can be accomplished by using a large carrier gas flow, operating the growth chamber at low pressure, or both. The distance to the growth zone should be kept as short as practical for purity considerations as well, as the purity of an undoped layer grown immediately after a heavily doped layer will be compromised if the undoped reactants must pass through a long length of tubing having walls that have just been saturated with dopant. This is particularly important for reactors using dopant precursors with serious memory effect problems, such as those for selenium and magnesium. However, if the common reactant path to the growth zone is too short, inadequate mixing may occur, resulting in poor crystal uniformity. In such cases, a mixing device can be placed in the gas stream to enhance the gas mixing.[55]

2.4 Reactor maintenance considerations

Not included in the simple reactor design was any consideration of the problems of reactor maintenace. Such consideration is essential if the reactor is to remain healthy and productive in the face of source changes and routine repairs. Most of these concerns are related to preventing or minimizing the exposure of internal reactor parts to air. Water vapor and oxygen react with many of the precursors used in MOCVD and both are extremely deleterious to the growth of AlGaAs. Water in particular clings tenaciously to stainless steel surfaces, and opening a section of tubing to air can result in a reactor recovery time on the order of days during which material quality is poor as this water is purged from the system. Baking air-exposed reactor parts can decrease this recovery time, but some parts such as mass flow controllers cannot be heated. It is thus critical that contact of internal reactor parts with air be minimized.

2.4.1 Component accessibility

All components should be easily accessible for ease of installation, removal, and leak checking. Convenience considerations aside, this is essential to prevent damage to easily damaged metal sealing surfaces while components are being handled inside the reactor. This also means that a component must be removable without having to remove other components. Removing a component exposes part of the reactor to air and results in a recovery time after the reactor is resealed during which the material quality is poor. Thus, components should be arranged such that opening one section of the reactor does not necessitate opening or otherwise contaminating unrelated sections. In addition, the plumbing should be compartmentalized with valves to as great an extent as possible so that maintenance work can be isolated.

2.4.2 Inert gas purge capability

In order to minimize air exposure when the reactor is opened, provisions should be made for an inert gas to reach all reactor plumbing. This inert gas should pressurize any section of line that is to be opened, so that the flow of inert gas out of the opened line will keep air out. This is particularly important for the line connecting the mixing manifold to the growth chamber which must be opened each time the chamber is loaded since exposure of residual reactants in this line to air can result in the formation of morphology-disturbing particles and in unacceptably high water and oxygen levels in the gas stream during growth. Suitable gases for this purpose are nitrogen, which is effectively inert for this application, and argon. Both are relatively inexpensive and easily obtained in high-purity grade and can be purified further with commercially available equipment. Helium is not suitable because residual helium left in the plumbing after purging interferes with the use of helium mass spectrometer leak detectors.

2.4.3 Evacuation and leak test capability

An important subsystem essential for the operation and routine maintenance that was not included in the simple reactor design is the plumbing necessary to evacuate and leak test all parts of the reactor. Central to such a system is a mechanical vacuum pump equipped with a durable, low vapor pressure oil. There are several synthetic oils available which are specifically formulated to withstand exposure to the chemicals used in the semiconductor industry and are particularly well suited for this application. Between the pump and the reactor should be a cold or sieve trap to prevent oil vapor from backstreaming into clean reactor areas during evacuation.[39] There should be an evacuation port on every subsystem so as to avoid having to evacuate one section of line through an unrelated section. In addition, there should be an evacuation port on the lines that connect the bubblers to the reactor so as to provide an evacuation path for these lines that does not contaminate other reactor plumbing with air after a bubbler change. Pumping through constrictions in the lines such as filters, mass flow controllers, needle valves and regulators is very slow and reduces leak detection sensitivity, and thus should be avoided. Components such as check valves, gas delivery pressure regulators and gas back pressure regulators are sometimes constructed such that they can only be pumped through in one direction, and even then may be impossible to evacuate completely, so a working knowledge of the internal construction of all reactor components is essential. The pumpout paths should be as short as possible to facilitate sensitive leak detection. The entire reactor must be extremely leak-tight. A maximum permissible leak rate of 10^{-9} standard cubic centimeters per second of helium is a limitation imposed by commercially available components. This level of leak detection requires the use of mass spectroscopic based equipment such as a helium leak detector.

2.4.4 Special purge considerations

Even if a reactor is leak tight there are still two sources of contamination with which to contend. First, it is virtually impossible to completely remove water from the internal stainless steel surfaces of the valves and tubing. This residual water is constantly outgassing into the reactant pathways. Second, since no seal is perfectly leak free, there is an extremely small but steady stream of air entering the reactor at all times. If allowed to accumulate, these contaminants will deleteriously affect the crystal quality until they are purged from the system. As a preventative measure it is advisable to constantly purge all reactor lines with high-purity hydrogen to prevent this accumulation. Toward this end, another modification to the simple reactor design is necessary. In the simple design, the only way to run hydrogen through a reactant line is to run it through a reactant bubbler. This is a waste of source material and a generally bad idea because after maintenance work the lines are likely to be contaminated, and this could contaminate the source or react with it. This problem is avoided by use of a source bypass configuration. Source bypass configurations for both bubbler and gaseous sources are shown in Fig. 12.

Another weakness in the simple design with regard to maintenance purging is the section of plumbing between a gaseous source tank and the main plumbing in the reactor. Because gaseous source tanks are usually located in exhausted cabinets separate from the main reactor plumbing, and sometimes even outside the building for safety reasons, there are long lines involved. As with the bubbler sources, purging these lines can only be accomplished by running the source itself through the line. Thus, high-purity hydrogen should be connected near the source tank to enable purging of these lines when not in use. This is also beneficial in that pumping through long lines at reduced pressures (<100µm) is extremely slow, so after maintenance work, air in such lines can be pumped out to a moderate vacuum and the remaining air can be removed by purging.

2.5 Component failure considerations

MOCVD reactors are complex machines that can be rendered completely useless by the failure of a single component. The most fortunate type of failure is one that is immediately identifiable and that has no long-term deleterious effects on the

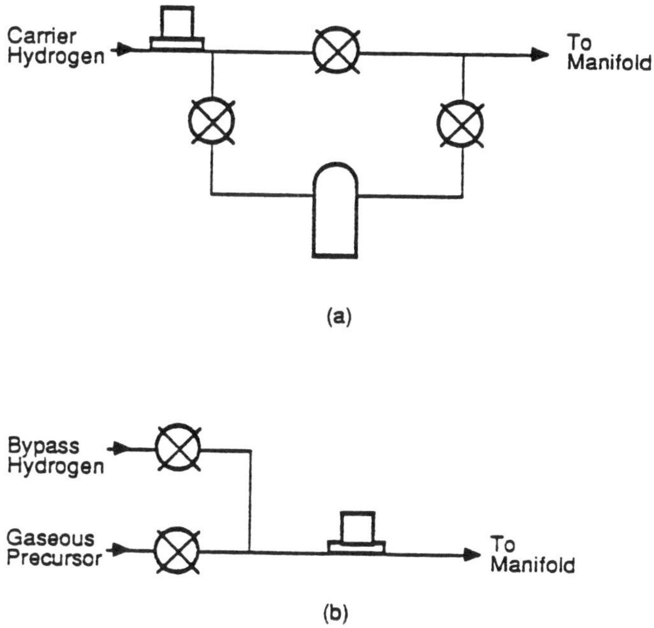

Figure 12. Source bypass configurations for (a) metal alkyl and (b) gaseous sources. For gaseous source lines the bypass hydrogen should be connected as close to the source tank as possible.

reactor after the component is repaired or replaced. Many component failures, however, are of a less fortunate type, being difficult to identify or damaging to the reactor, or both. Some component failure modes and design strategies to minimize their impact are discussed below.

2.5.1 Leaks

The one failure mode that is common to all components is leakage to air around seals and system connections. An air leak, depending on its size and location, can cause extremely subtle changes in material quality, or it can cause sudden changes in crystal doping efficiency, surface morphology and device performance. In general, a leak from a line at greater than atmospheric pressure, which tends to expel gas, is less obvious in terms of reactor performance than one from a line at or below atmospheric pressure, which tends to allow air into the system. Of course, any leak is a possible health and safety hazard, and for this reason as many potential leak points as possible should be enclosed in exhausted cabinets for containment. Gas monitors should be thoroughly distributed to aid in identifying the presence of and locating leaks. Some components, such as regulators and bellows valves, have special ports connected to the external volumes around seals that can be plumbed into a leak detection system to provide an immediate indication and location of a leak. Placement of the reactor in an environment free of vibration and shock can aid in preventing leaks.

2.5.2 Gas regulators

A gas regulator can fail in several ways. It can simply function poorly by giving unacceptably large variations in output pressure for different input pressures or output flows, or by not regulating output pressure within its specified range. Such failures are easily identified and rectified. A far more critical failure is one in which the regulator set point shifts to give an output pressure much higher than intended, in worst case equal to the regulator input pressure. If the increased pressure is much greater than approximately one hundred pounds per square inch, then the normally-closed air actuated bellows valves that are typically used will be forced open. This is not damaging to the valves, but it will allow the high pressure to reach the mass flow controllers, which may be permanently damaged by such high pressure and develop a leak to the atmosphere. Additionally, the hydrogen purifier is particularly susceptible to damage from excessive input pressure. As a precautionary measure, there should be a pressure sensor system on the output side of each regulator to shut off the gas feed and sound an alarm in the event of an over-pressure condition. There should also be a pressure relief valve downstream of each regulator, plumbed to send released gas to vent.

2.5.3 Bellows valves

As with other types of valves, bellows valves can be operated only a finite number of cycles, typically in the hundreds of thousands range, before a leakage path develops between the in and out ports. This lifetime, however, can be substantially reduced by the presence of the particles on internal sealing surfaces when metal alkyls come into contact with minute quantities of air.[19] Such particles can become imbedded in the soft sealing surface of the valve and prevent the stem tip from solidly contacting and sealing against the seat. Thus, for valve leakage considerations alone it is critical that the reactor be leak-tight and that all maintenance operations that involve exposure of the internal reactor surfaces to air be preceded and followed by thorough pumping, purging and baking to eliminate traces of metal alkyls and air that could contribute to the formation of these particles. As a practical matter, though, it is impossible to prevent their formation in an active reactor since leaks develop and maintenance accidents occur. The probability of causing leaks, however, can be minimized by careful valve orientation. In Fig. 13 is shown a cross-sectional view of an

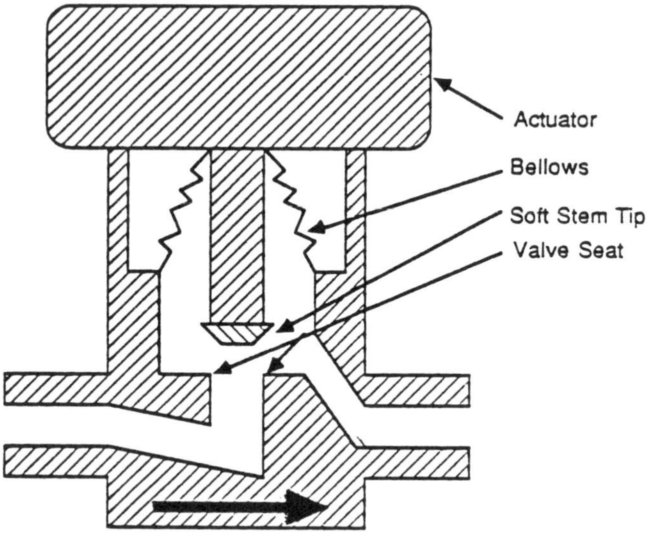

Figure 13. Cross section of a Nupro Company BK bellows valve. The arrow on the valve indicates the manufacturer's suggested gas flow direction.

air actuated bellows valve in the open position. Note the asymmetry of its configuration. The bellows is downstream of the valve seat if the gas flow is orientated as indicated by the arrow on the valve body. This orientation is recommended by the manufacturer because the valve will close against a slightly higher upstream pressure than if it is oriented oppositely, although otherwise the valve will function equally well in either orientation. For pressures likely to be encountered in a reactor, typically 20 pounds per square inch or less, this is not a factor, and the reactor designer is free to orient the valve in either direction. Consider a situation in which the valve is between a reactant line and vent line with the suggested valve orientation, i.e., with flow as indicated by the valve arrow. If with this valve closed the vent line is exposed to air because of a leak or maintenance operation, the internal surface of the valve bellows will also be exposed to air. This could lead to particle formation on the bellows surface due to metal alkyl traces present either before or after the air exposure. If the particulates had formed on a rigid surface such as a tube wall, this might not be a problem at all since they would be not be free to move and end up in the valve seal. However, because the bellows flexes with valve operation, particulates there will be dislodged and, if the valve is oriented with the actuator up, fall into the valve seal area, thus greatly increasing the possibility for valve leakage. This gives two design rules concerning bellows valves. First, valves that are exposed to reactants should be installed with the actuator down, and second, valves should be oriented such that when the valve is closed the bellows faces the side of the line that is less likely to be exposed to contaminants, usually the upstream side. This is particularly important for valves that might occasionally be exposed to air such as those that control flow to the reaction chamber.

Another possible failure mode is, for pneumatically actuated bellows valves, malfunction of the valve actuator system which could result in a valve being in an unexpected state. This consists of the valve actuator itself, a separate electrically operated solenoid to direct compressed gas to the valve and the compressed gas source, usually an air or nitrogen tank. The simplest failure of this system, loss of gas pressure, should be anticipated with a back-up source that automatically switches in when the pressure drops. The solenoid is susceptible to an interruption of electric service, and this will be discussed in a later section. Both the solenoid and the valve actuator are susceptible to leaking seals that can reduce gas pressure in the line to the actuator to the point where the valve only opens (or closes, if of the normally open type) partially. Such leakage cannot be directly monitored short of putting a pressure sensor on each valve actuator, which is hardly practical, or by monitoring the usage rate of the compressed gas and noting sudden increases. Some bellows valves are available with an electrical switch that is mechanically linked to the actuator in such a way that the state of the switch indicates the state of the valve. Although this might not detect a partially opened or closed state, it would indicate a total failure of any of the components involved in the actuator system.

2.5.4 Hydrogen purifier

The hydrogen purifier, as shown schematically in Fig. 14, is essentially a hot palladium sheet through which only hydrogen can diffuse, leaving behind the impurities present in the feed gas to be swept out with the bleed flow. If the palladium cools to near room temperature from its operating temperature of typically 700° F, either due to an elecrical service interruption or a failure of the temperature controller, there are two problems. First, the hydrogen output from the purifier will drop to zero, causing gas flow in the reactor to stop and allowing any reactants present in the lines to diffuse freely about the reactor through any open valves. This problem is compounded if the lack of flow allows air to enter the reactor through vent lines and react with the metal alkyls present. This situation should be prevented by use of a system to detect a drop in purifier delivery pressure below some minimum value,

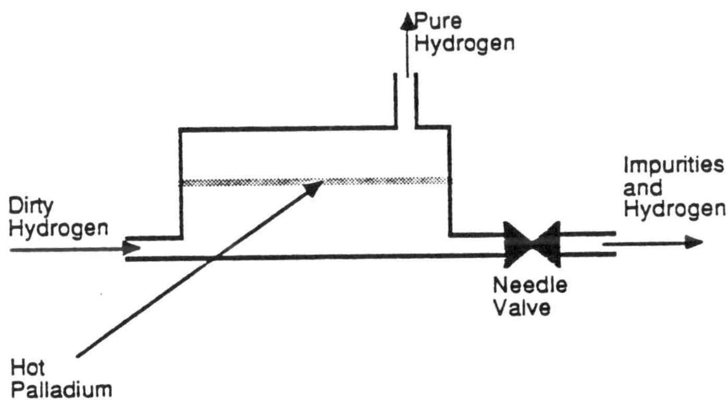

Figure 14. Cross section of a hydrogen purifier.

activate an alarm and then switch in an independent high-purity gas source, such as hydrogen, argon or nitrogen, in order to purge the reactor prior to shutdown. In addition to the problems of loss of hydrogen supply, an unplanned purifier cool down is potentially damaging to the purifier itself. If the cooling takes place in the presence of hydrogen, as is likely in an unplanned cool down, then hydrogen atoms traveling through the palladium will be frozen in place as the palladium cools and contracts. The stress induced by the presence of the hydrogen is likely to cause microcracks in the palladium and result in a leakage path from the dirty input hydrogen side to the clean output side. If operated with such cracks, the purifier will contaminate the reactor with dirty hydrogen, with the degree of damage thus sustained will depend upon the size of the leak and the amount of time before it is detected. Therefore, the purifier should be checked for leaks at operating temperature and pressure after every unplanned cool down.

The hydrogen purifier is also susceptible to damage from excessive input hydrogen pressure. Because the output of purified hydrogen is a strong function of the input pressure, purifiers are frequently operated at maximum input pressure in order to obtain the maximum delivery rate of pure gas. However, this means that a small increase in delivery gas pressure due to a regulator malfunction can cause the pressure to exceed the damage threshold, resulting in damage to the seals between the input and output sides of the purifier, with the same potential for reactor damage as mentioned above. Unfortunately, there will very likely be no external indication of such a leak until after the damage is done. Thus, continuous monitoring of the quality of the purifier output is necessary to warn of such degradation before significant damage can occur. The main impurities present in the low-grade hydrogen typically used for purifier feed are the components of air, mainly oxygen, nitrogen and water. The purifier output should be monitored for any of these as a purifier leakage indication. The detection of a leak should result in alarm activation and the automatic replacement of purifier hydrogen to the reactor with an alternate high-purity gas source to purge the reactor prior to shutdown.

2.5.5 Mass flow controllers

A mass flow controller, like a gas regulator, is relatively limited in its failure modes. It can control poorly by giving unstable flows about its set point, not control at all by giving a flow that is independent of its set point or lose its calibration. Unlike a gas regulator, however, a mass flow controller failure has no potential for reactor damage, although it can obviously affect the composition of the gas stream during crystal growth. Precautions must be taken that ensure the detection of an intermittent failure that might go unnoticed by casual observation. Undetected intermittent flow variations can cause subtle variations in crystal parameters and resulting device performance that might otherwise be attributed to other problems with the reactor. Thus, mass flow controller performance should be continuously monitored by comparing the actual flow to the set point, and an alarm should be activated when a flow is deemed to be out of acceptable range. This is easily done with computer monitoring.

Most commercially available mass flow controllers have an integral flow sensor that works by diverting a small fraction of the flow through the controller through a thin tube. The temperature of the gas is measured on entry. It is then resistively heated and its temperature is measured once again before it rejoins the main stream. The flow rate is a function of the temperature difference.[56] Problems may arise if the gas in the stream is thermally unstable enough to decompose when heated in the sensor tube and result in partial clogging of the tube, thus altering the flow sensor calibration. For this reason, and to avoid exposing the rubber seals in the controller to reactants, mass flow controllers are usually placed upstream of the metal alkyl sources. For situations where the reactant must pass through the controller, such as a controller on an arsine line, the problem can be reduced by using diluted rather than pure gas in the system.

2.6 Human interface considerations

Even if a reactor is well engineered, operator error can cause serious damage to the system. Through the use of computer control and monitoring, the role of the operator in reactor operation can be reduced nearly to loading and unloading crystals. A computer can operate all valves and monitor all process parameters. It can also monitor alarms and take preprogrammed actions in response to various trouble indications. This is particularly useful in panic situations, such as toxic gas leak alarms, when an operator cannot stay in the reactor area, and might anyway be too upset to correctly identify the problem and take action to neutralize it in a manner that minimizes possible safety hazards and reactor damage. However, it would be difficult to anticipate in the computer programming every possible failure and necessary response, so a knowledgeable operator would still be needed. Operator errors can also occur during maintenance operations such as oversights in proper valve operation order when evacuating or pressuring lines. Such errors might be avoided by having the state of each valve, including those that are hand operated, automatically indicated on an easily interpreted status board.

3. Reactor Control Software

Although it is not an absolute necessity for simple device structures, computer control of an MOCVD reactor is critical to reproducible thin layer growth and to enable the growth of graded structures that require the accurate and continuous variation of multiple growth parameters. The advent of inexpensive, sophisticated microcomputers and related equipment for real-time control applications has made reactor computer control available at a small fraction, in terms of both expense and effort, of total reactor cost and thus it is the rule rather than the exception. Computer control capability, however, is limited by the capability of the computer software. In this chapter, some reactor control software will be described.

3.1 Control software organization

In general, a reactor control software system is composed of two main modules. The first module, the control module, is a computer program that handles timing, status display and the communications necessary with interface hardware such as analog-to-digital and digital-to-analog converters and control pulse routers, making such actions transparent to the reactor user. This module also performs autonomous background functions such as the monitoring of flows and alarm indicators. The second module, the growth program module, conveys information to the control module concerning valve and flow settings and growth times for each layer for a particular growth, and thus is unique to a particular crystal structure. This module may be an actual computer program that calls the control module as a subroutine for each layer to be grown, or it may simply be a list of parameters that the control module reads as needed. There are positive and negative aspects to both approaches. There can be a significant savings of execution time for the latter case, as all communication between the growth module and control module can be completed before the growth actually starts, thus saving the time associated with a subroutine call prior to the growth of each layer. However, this approach also limits flexibility in that it requires that the control module contain all the code required to grow every type of layer that might be desired, so that a layer calling for an unusual grading function or valve configuration necessitates the changing of the control module. The control module, because of its many functions, tends to be complex, so the likelihood of error in its modification is rather high. Also, with each section of rarely used custom code added, the control module becomes larger and thus requires more computer memory. The other approach has the advantage of flexibility since the growth program can be made as complex as necessary for unusual device structures without disturbing the control module. An example of such a device is the CHIRP (coherent interfaces for reflection and penetration) diode,[57] which is a superlattice in which layer thicknesses are adjusted to give a linearly varying miniband gap in order to realize negative differential resistance in the dc I-V characteristic. In a research environment where a variety of novel structures is likely to be grown, the user-written computer program growth module approach is preferred. This approach is the one taken on the reactor used in this work. Briefly, the system consists of a compiled BASIC language control module that performs all the hardware communications mentioned above, and a BASIC language growth program that specifies parameters for each layer and calls the control module as a subroutine. When it receives control from the growth program the control module implements the hardware configuration called for and, while awaiting the completion of the layer being grown, obtains the parameters for the next layer and decodes them, thus allowing an instant switch to the next layer upon completion of the layer in progress. Layers with as short a growth time as one second can be grown with this system. The control hardware consists of an Apple II+ computer equipped with a Thunderclock clock board, and a Digital Equipment Corporation BIN III data acquisition and control system. A block diagram of the system is shown in Fig. 15.

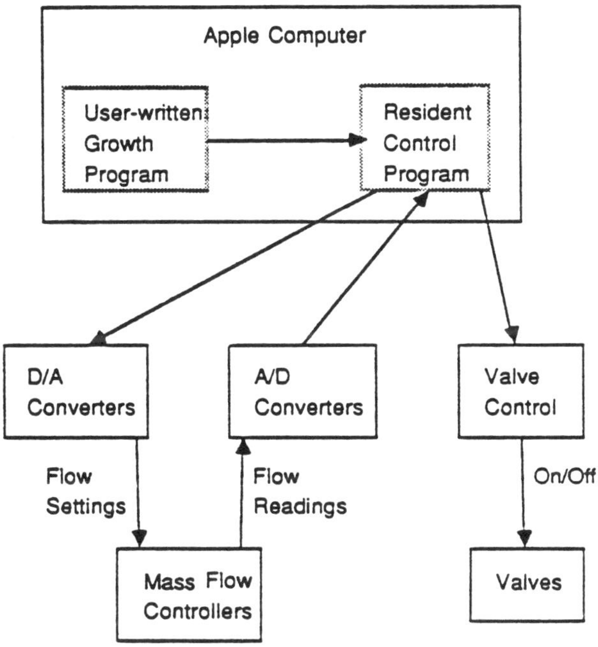

Figure 15. Block diagram of the reactor control system.

3.2 Thin layer growth algorithm

The controllable and reproducible growth of thin layers by MOCVD is possible through precise control of growth times and reactant flows. The computer control system implemented on the reactor used in this work allows layer growth times that are an integral number of seconds long. This is not a limitation of the control system, but rather a limitation imposed by the fact that it takes several hundred milliseconds for a pneumatically actuated bellows valve to open or close. Finer layer growth time resolution would be of dubious usefulness because of the uncertainty due to valve operation transients. In order to minimize the impact of such transients, the layer growth time should be long compared to the valve operation time. With regard to flow controllability, the Tylan mass flow controllers used on this reactor allow reactant flow control to better than 1% of the set point over the range of 100% to approximately 10% of the controller full-scale flow. Below 10%, the actual flow is a nonlinear function of the set point with the exact relationship unique to a given mass flow controller. In the 2 to 3% of full-scale range, the flow may vary from the set point by as much as 50% and have an oscillatory component that is a significant fraction of its desired value. Closure below 2% of full scale is generally not possible.

These limitations must be taken into account when determining reactant flows and growth times for thin layer growth. In order to define growth parameters for a layer of arbitrary thickness and composition, the following must be taken into account simultaneously: the allowable flow ranges for the TMGa and TMAl mass flow controllers taking into account the nonlinearities mentioned above, the relationship between AlAs mole fraction and the TMAl to TMGa flow rate ratio and the AlAs and GaAs growth rates per unit flow rate of TMAl and TMGa, respectively. This is a tedious, time-consuming process that is well suited to implementation in a computer program. Such a program will now be described.

The user is first prompted for the normalized growth rates of GaAs and AlAs in units of microns per minute per sccm of reactant flow, and then for the allowable maximum and minimum flows for the TMGa and TMAl mass flow controllers. Finally, the user is prompted for the thickness and AlAs mole fraction (x of $Al_xGa_{1-x}As$) of the layer to be grown. No further input is required. The program first determines the TMGa and TMAl flows, within the boundaries specified, that will give the specified layer composition at the minimum growth rate. If the layer is GaAs or AlAs, then only the TMGA or TMAl flow, respectively, will be switched into the growth chamber, and the other flow is irrelevant. If the layer is AlGaAs, then the necessary flow ratio is calculated by use of the relation

$$F_{Al}/F_{Ga} = 2x/(1-x) \tag{3.1}$$

where F_{Al} is the TMAl flow, F_{Ga} is the TMGa flow and x is the mole fraction of AlAs. This relation is unique to the TMGa and TMAl bubbler temperatures for the reactor used in this work. Note that the flows are equal for $x=1/3$, the TMAl flow is smaller for $x<1/3$ and the TMGA flow is smaller for $x>1/3$. Also note that imposing limits on mass flow controller settings also directly defines a range of achievable x. The smaller of the TMGa and TMAl flows for the desired composition is thus determined and set to its lower bound, and the other flow is calculated using Eq. (3.1). The minimum growth rate R is then given by

$$R = F_{Ga} * A + F_{Al} * B \tag{3.2}$$

where A and B are the normalized GaAs and AlAs growth rates, respectively, and F_{Ga} and F_{Al} in this case are the minimum TMGa and TMAl flows, respectively, to give the desired composition. The maximum growth rate R_{max} is determined in an analogous manner, with the larger of the two flows set to its upper bound and the other flow determined from Eq. (3.1). The minimum and maximum growth times t_{min} and t_{max}, respectively, are then determined by dividing the desired layer thickness by R_{max} and R_{min}, respectively. These growth time bounds will in general not be integers. Valid growth times are the integers between the bounds. At this point it is a simple matter to determine for each possible growth time the growth rate, and thus the TMGa and TMAl flows, necessary to give exactly the desired layer thickness and composition. It is possible, if the layer is extremely thin and the limits on the mass flow controllers are too restrictive, that there are no integers between the bounds. In that case, the limits must be relaxed or a compromise in layer thickness must be accepted. Such a compromise in layer thickness can be generated by including in the range of growth times t_{min} rounded down and t_{max} rounded up, and then multiplying them by R_{max} and R_{min}, respectively, to give layer thickness options above and below the target thickness.

An example of the output for this program is shown in Fig. 16. All possible growth times are shown, and it is up to the user to choose one that is suitable. The chosen growth time and reactant flows are then inserted into the growth program module. As discussed above, the growth time should be long enough such that the effects of switching transients are minimized. A minimum of several seconds has permitted reproducible thin layer growth in this laboratory. Another possible consideration may be the need to grow the layer at reactant flows similar to those used previously for bulk doping check layers, which would add constraints to the choice of flow ranges and growth times.

3.3 Graded layer growth algorithm

The growth of layers graded in composition or doping is generally accomplished in MOCVD by continuous variation of the reactant gas phase stoichiom-

```
Growth rate A = .0036
Growth rate B = .0018
Min and max Ga flows = 8,20
Min and max Al  flows = 8,20
Description for this layer:
? 100A undoped 30% AlGaAs
What is X? .3
Thickness (Å)? 100
Doping type? (N,P,U) U
Possibilities:

seconds    L (Å)    Fga      Fal

   6       100      19.44    16.66
   7       100      16.66    14.28
   8       100      14.58    12.50
   9       100      12.96    11.11
  10       100      11.66    10.00
  11       100      10.60    9.090
  12       100      9.722    8.333
  13       104      9.333    8.000
```

Figure 16. Example session of the thin layer growth set-up program User input is underlined. The output consists of TMGa and TMAl flow combinations and growth times that will produce the specified layer.

etry in the growth chamber during the growth. This is easily done by varying reactant mass flow controller settings. There are several approaches to the organization of the control software necessary for this task. The actual mass flow controller settings must be varied during the growth, and this requires that the grading software execute quickly. Since the actual variation of flows will be done quasicontinously, i.e., in short time steps of constant flows, it is desirable to make the time steps as short as possible in order to approximate the desired grade as closely as possible. If the grading software executes slowly, then it, rather than the system time overhead, might be the limiting factor in the determination of the minimum possible time step. If the grading routine is designed such that the reactant flows for a given time step are calculated during the growth of the preceding time step, then the lower limit on the time step size is equal to the execution time of the flow determination routine. Depending upon the complexity of the grading function and flow selection algorithm, this could be an unacceptably large time with the resulting grade being a poor representation of the intended profile. One solution is to calculate the flows required for each time step, save them before the growth is started and read them as needed during the graded layer growth, thus saving calculation time during the growth and enabling overhead-limited growth step times. This approach can exact a high price in computer memory, however, especially if there are several graded layers in a single reactor run. Still, it is the fastest possible technique. As a practical matter, present-day microcomputers have adequate computational speed to permit relatively complex grading algorithms to be executed quickly enough that step time restrictions are due to other system time bottlenecks such as interface hardware reading, writing and error checking. Thus it is possible and desirable, in keeping with the goal, to place as much of the sophistication as possible in the growth program for flexibility and easy modification, to calculate flows for each time step in real-time.

The control system for the reactor used in this work allows the growth of layers as short as one second in duration. This is the system overhead limited time step length. Due to control system timing restrictions, the reactant flows cannot be read and displayed during a layer this short. Two seconds is the minimum growth time that will permit this. While it is not essential that the reactant flows be monitored since the layer growth will proceed regardless, it is reassuring to see that the flows are accurately tracking the rapidly changing set points. The grading routine developed for this reactor allows the growth of layers graded in doping or composition, or both.

3.3.1 Growth of layers graded in doping

In order to enable the growth of the full range of device structures otherwise possible by MOCVD, the doping system must be capable of producing controllable doping over a range of approximately three orders of magnitude, approximately from 10^{16} to 10^{19} cm^{-3}. This could be accomplished simply by varying the dopant flow from a gaseous dopant source, or the carrier gas through a metal alkyl dopant source, over a similar range. Unfortunately, commercially available mass flow controllers can regulate flows over a range of less than two orders of magnitude, and as discussed in Section 3.2, part of this range may not be useful. In order to achieve the necessary range of doping, a dilution system is used. Such a system is schematically shown in Fig. 17. This system consists of three mass flow controllers: one for the dopant source, one for dilution and one to control the amount of diluted dopant sent to the reaction chamber. A back pressure regulator is placed at the junction of the three controllers to hold that point at a constant pressure regardless of the individual flow settings for stable mass flow controller operation and to vent any diluted dopant precursor that is not sent to the reaction chamber. The dilution is described mathematically as

$$F_{net} = F_{tot} * F_{dop} / (F_{dop} + F_{dil}) \quad (3.3)$$

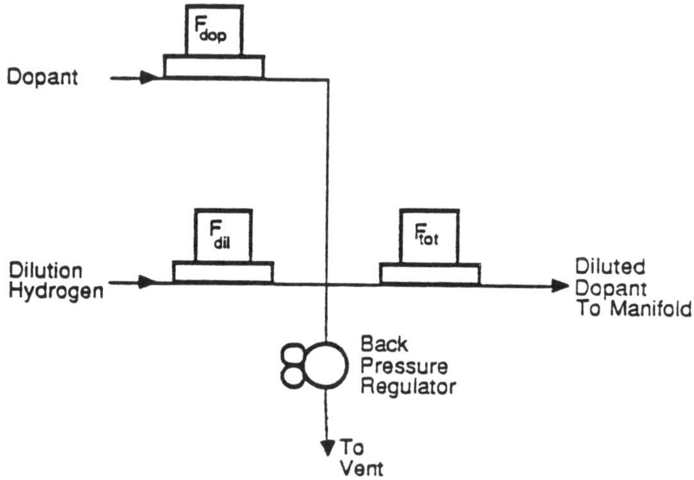

Figure 17. Schematic of a dopant dilution system.

where F_{dop} is the flow rate of dopant in carrier gas, F_{dil} is the flow rate of the dilution gas, F_{tot} is the flow rate of diluted dopant into the reaction chamber and F_{net} is the equivalent undiluted source flow that would be necessary in order to give the same dopant delivery rate as that achieved with the dilution system. F_{net} describes the actual dopant precursor flow into the reaction chamber and hence the crystal doping. Note that F_{net} is a nonlinear function of both F_{dop} and F_{dil}, and is a linear function of F_{tot}. On the reactor used in this work the full scale flows of F_{dop}, F_{dil} and F_{tot} are 100, 1000 and 100 sccm, respectively. Assuming a well-behaved flow range from 10 to 100% of full scale for each controller, the range of F_{net} is from 0.1 to 50 sccm, which is not quite three orders of magnitude. However, if a range from 5 to 100% of full scale is permitted, then the range of achievable F_{net} becomes 0.025 to 50 sccm, which exceeds three orders of magnitude.

The dopant dilution system presents a problem with regard to the growth of a layer graded in doping. Consider a situation in which a doping grade is performed by varying only F_{tot} of Eq. (3.3). Because of the linear dependence of F_{net} on F_{tot}, this is a convenient and logical approach. In this case, referring to Fig. 17, the section of gas line between the F_{tot} mass flow controller and the manifold contains a constant concentration of dopant since F_{dop} and F_{dil} are never changed. When F_{tot} is changed, only the velocity and hence the delivery rate into the manifold of the dopant gas are changed. Assuming negligible pressure gradient in that section of line, a change in F_{tot} will produce an instant change in the gas velocity throughout that line and thus an instant change of the dopant delivery rate into the manifold. There is no time lag involved. Now consider the alternate situation in which a grade is performed by varying either F_{dop} or F_{dil} while F_{tot} is held constant. Here, from one time step to the next the concentration of dopant on the input side of the F_{tot} mass flow controller changes. This new concentration must traverse the section of tubing between this mass flow controller and the manifold before the change in F_{dop} or F_{dil} is seen by the manifold. Since the gas velocity in 1/4-inch outside diameter tubing is about 0.1 inch per sccm per second, there is likely to be a delay of several seconds between the dopant dilution change and its manifestation at the manifold. This is a difficult effect to account for in a grading routine, especially since in addition to the delay there will be a mixing transient in the section of the plumbing where the dilution takes place. Thus, the only practical method of grading doping in this system is by the variation of the F_{tot} flow. This limits the grading range to the range of the F_{tot} mass flow controller, which is less than two orders of magnitude for a single layer. Greater range can be achieved by dividing the layer into smaller layers and using suitable F_{dop} and F_{dil} flows for each section to allow access to the necessary doping range by the variation F_{tot}.

The doping grade routine developed for the system used in this work allows linear and parabolic grades of the F_{tot} flow for a layer. For a layer of constant alloy composition, this translates directly into linear and parabolic profiles of dopant incorporation into the layer. It should be noted that this does not necessarily translate directly into free carrier concentration due to finite impurity ionization energy, nonzero background doping concentration and heavy doping effects.[58] Also, for a compositionally graded layer, the same caveats apply, and in addition the variation of impurity ionization energy[59] and background doping concentration with composition[13] must also be taken into account when attempting to implement a specific carrier profile.

3.3.2 Growth of layers graded in composition

The growth of a compositionally graded $Al_xGa_{1-x}As$ layer ($0<x<1$) is accomplished through a quasicontinuous variation of the TMGa and TMAl flows into the reaction chamber in such a way as to produce the desired compositional profile. The composition grown during any given step is determined by the ratio of these flows and is given by

$$x = (F_{al}/F_{ga})/(2+F_{al}/F_{ga}) \tag{3.4}$$

which is an inversion of Eq. (3.1). Thus, the task of the grading routine is to determine F_{ga} and F_{al} for each time step. There are, however, several conditions that the selected flows must meet. As discussed previously, mass flow controllers have a limited useful flow range that must be taken into account since this limits the achievable flow ratio and thus the compositional range. In Fig. 18 is shown the flow ratios necessary to achieve the compositional range $0.01 \leq x \leq 0.96$. As can be seen, extremely low and high compositions require very high flow ratios. The TMGa and TMAl mass flow controllers on the reactor used in this work both have 100 sccm full scale flows. Restricting their operation to the linear range of 10 to 100% of full scale limits the composition to the range $0.05<x<0.83$. Relaxing the lower limit of this range to 5% of full scale increases the range to $0.02<x<0.91$ at the expense of accuracy in the added range.

This brings up the flow selection criterion that the crystal growth rate defined by the chosen flows must be small enough to ensure good crystal quality and to permit each individual step in the grade to be small compared to the overall graded layer length. In Fig. 19 is shown the relationship between minimum growth rate and composition for several different mass flow controller ranges. As can be seen, for a given set of ranges the growth rate becomes largest near the extremes of composition and is especially large for the ranges with large upper bounds. Thus, there is a conflict in requirements as the mass flow controller ranges that allow the greatest range of achievable compositions also permit the largest growth rates and thus result in the coarsest grades.

The final criterion for the selection of flows is that the flow increments from step to step must be very small compared to the full-scale flow. This is necessary in order to ensure that the mass flow controllers track the changing set points without large oscillations. For example, if a flow that is set at 20 sccm for one time step is decreased

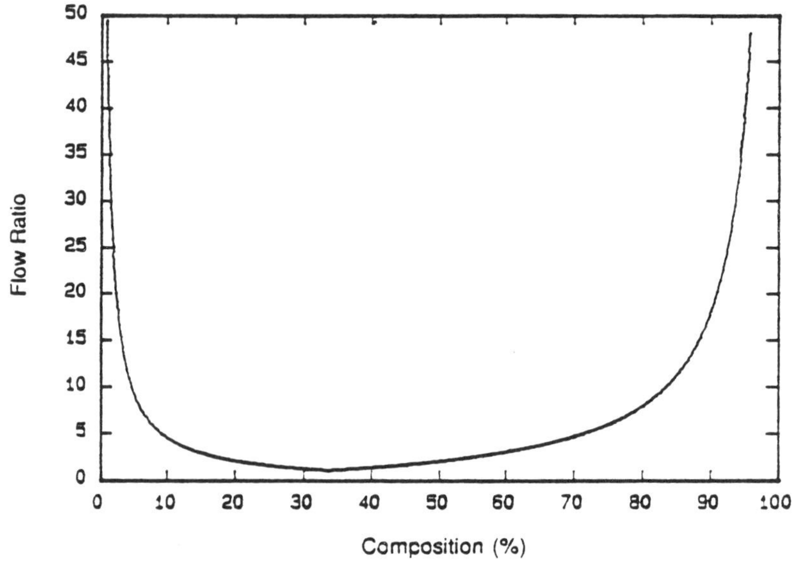

Figure 18. Flow ratio versus aluminum composition x. For x<1/3 Fga/Fal is shown, and for x>1/3 Fal/Fga is shown. The ratio is unity for x=1/3.

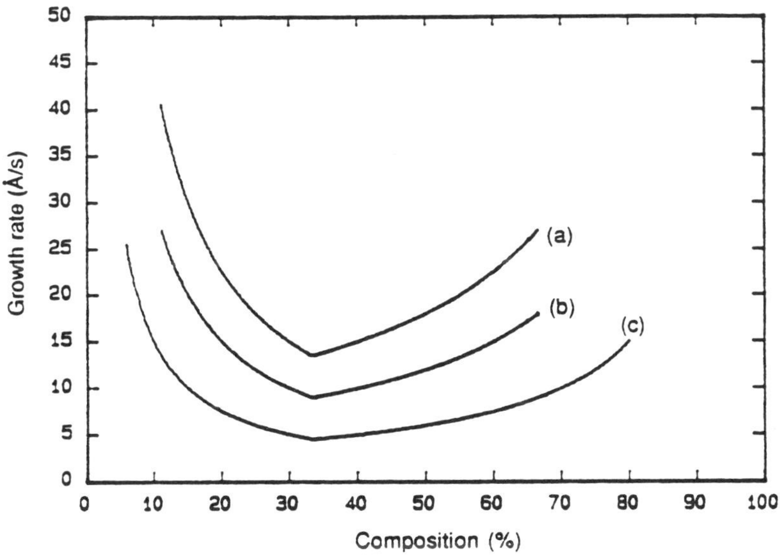

Figure 19. Minimum growth rates versus composition for several mass flow controller ranges. Minimum and maximum flows in sccm for both TMGA and TMAl are (a) 15, 60, (b) 10, 40 and (c) 5, 40. For this figure the normalized growth rates for GaAs and AlAs were taken as 0.0036 and 0.0018 microns per minute per sccm, respectively.

to 10 sccm for the next time step, there will be a settling time of several seconds as the flow stabilizes. This settling time may exceed the time step size, resulting in a significant deviation from the desired compositional profile.

The compositional grading algorithm developed requires as input the upper and lower limits for the TMGa and TMAl flows, the normalized growth rates for GaAs and AlAs (A and B of Eq. (3.2)), the total length of the layer, and information to define the variation of the desired compositional profile with distance from the start of the layer. For a linear grade the starting and ending compositions are required, and for a parabolic grade a third location and composition are required as well. The boundary conditions are used to determine the coefficients of the general parabolic equation

$$x = z\,(az + b) + c \tag{3.4}$$

where x is the desired AlAs mole fraction at distance z from the start of the graded layer and a, b and c are boundary-dependent coefficients. Equation (3.4) is written in this computationally efficient form to save computer execution time. Other grading functions can be substituted as needed. When a graded layer is to be grown, the graded layer control routine, located at the end of the growth program, is called with the input data described above. Program control remains in this routine until the graded layer is completed.

As mentioned before, the grading routine must resolve the conflict between the requirements for a wide achievable composition range and low growth rates. This is done by selecting the minimum flows, and thus the lowest growth rate, for each step that will produce the desired composition for that step. It is possible, however, that the desired composition cannot be grown within the imposed flow limits, so the imposed limits must then be violated. However, the rule observed by the algorithm in such a case is that a maximum flow limit may never be exceeded, as this would result in a growth rate larger than the limit implicitly specified by the flow limits. Thus, in order to achieve the desired composition the larger of the flows is set to its specified maximum, and Eq. (3.1) is used to determine the other flow, which will then be less than its specified minimum. Due to mass flow controller nonlinearity and instability in the lower ranges, this may result in some inaccuracy in the grown composition. If this is unacceptable for the structure being grown, then the upper bounds on the reactant flows must be increased. It is up to the crystal grower to balance these considerations for the crystal structure being grown through the choice of flow limits, and in fact flow limits may deliberately be chosen such that they will be violated as described above in order to avoid excessive growth rates at the extremes of composition. The only limitation on allowable flow ranges imposed by the grading algorithm is that the minimum limits for the TMAl and TMGa flows must be identical. The algorithm will fail to consistently generate flow settings that are within the specified limits if the minimum flows are not equal. A flowchart of the grading algorithm is given in Fig. 20.

As a detailed example of the operation of the grading routine, consider the growth of a parabolically graded layer 0.12 μm thick that starts with $x=0.2$, ends with $x=0.85$ and has $x=0.3625$ at 0.06 μm from the layer start. Such a grade is used in the graded index quantum well laser diodes[59] having $x=0.85$ confining layers grown in this laboratory. Referring to Eq. (3.4), this gives $a=45.1388889$, $b=0$ and $c=0.2$. The normalized GaAs and AlAs growth rates are, referring to Eq. (3.2), $A=0.0036$ and $B=0.0018$ μm per minute per sccm, respectively. Finally, the flow limits are 10 and 20 sccm for TMGa and 10 and 31 for TMAl. The graded layer control subroutine is called with these input data. The subroutine first calculates the minimum and maximum compositions possible with the specified flow limits, which for these limits are $x_{min}=0.2$ and $x_{max}=0.63$ and then sets the "graded layer thickness grown thus far" variable z to zero. Next, a loop is entered that will be repeated once for each time step until the layer is complete. At the beginning of this loop the composition desired for the present

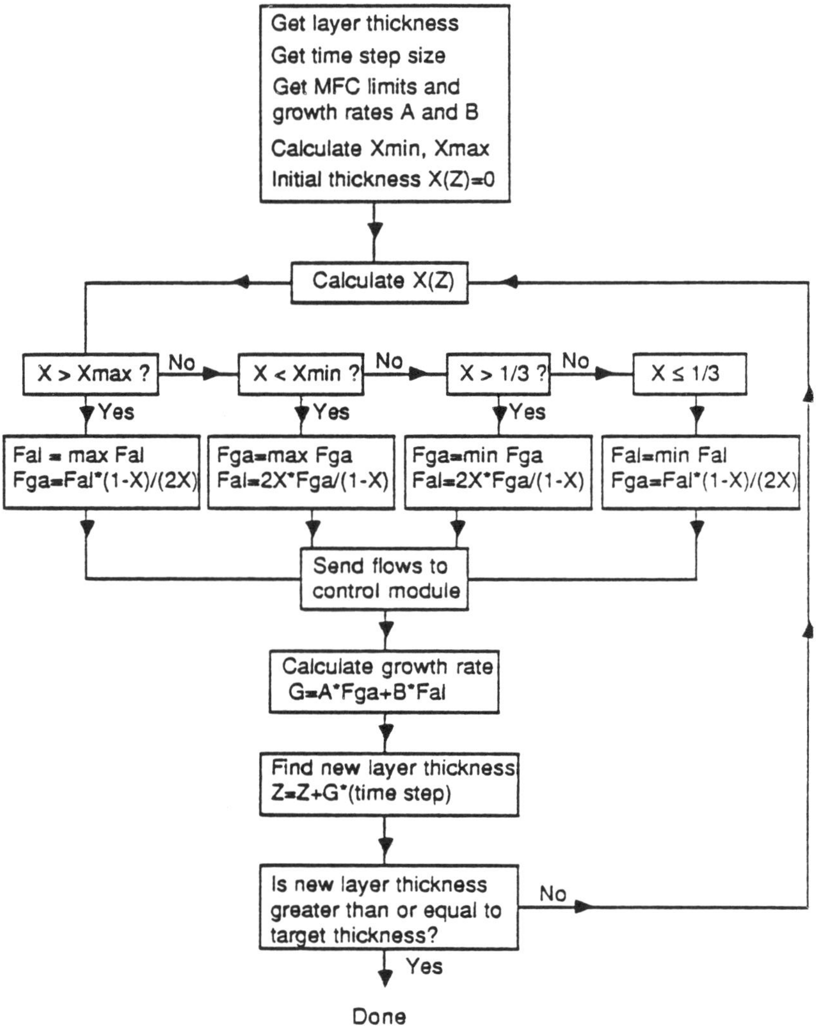

Figure 20. Flowchart for the compositional grading algorithm.

location in the layer is calculated. For z=0 it is found that x=0.2. This composition is compared to x_{min} and x_{max} and found to be achievable without violating the specified flow limits. For this composition (and all x<1/3) the TMAl flow is smaller than the TMGa flow, so the TMAl flow is set to its minimum value of 10 sccm, and Eq. (3.1) is used to determine the necessary TMGa flow which is found to be 20 sccm. Next, the control module subroutine is called to set the flows as calculated and initiate the growth of the layer. While that section of the layer is growing, Eq. (3.2) is used to determine that the growth rate is 15 Å/s, and this multiplied by the time step of two seconds gives a thickness contribution of 30 Å by the segment being grown. This is added to the thickness counter z to yield a total thickness grown of 30 Å after the present time step, and this is compared to the target thickness of 0.12 µm. Since the total thickness grown is less than the target thickness, control is returned to the top of the loop, the updated z is used to determine the composition of the next step and appropriate flows are determined. These calculations are performed during the growth of the previously calculated layer, and when that time step is complete the next segment of the layer is started with no interruption in the growth. This process is repeated until the total grown thickness is greater than or equal to the target thickness, and then the grading subroutine returns control to the growth program for the growth of the next layer. Figure 21 shows the compositional profile and variation of TMGa and TMAl flows throughout this layer, assuming that the mass flow controllers give exactly their set flows. The change of the smaller flow from TMAl to TMGa can be seen at x=1/3, and the deliberate exceeding of the lower bound of the TMGa flow can be seen for compositions greater than x=0.63, the maximum composition with the specified flow bounds. Figure 22 shows the algorithm-generated compositional profile and reactant flows for the same layer and flow bounds except the time step size is five seconds instead of two seconds. This grade is significantly coarser than that obtained with the two-second time step. Also note that toward the end of the layer the TMAl flow incre-

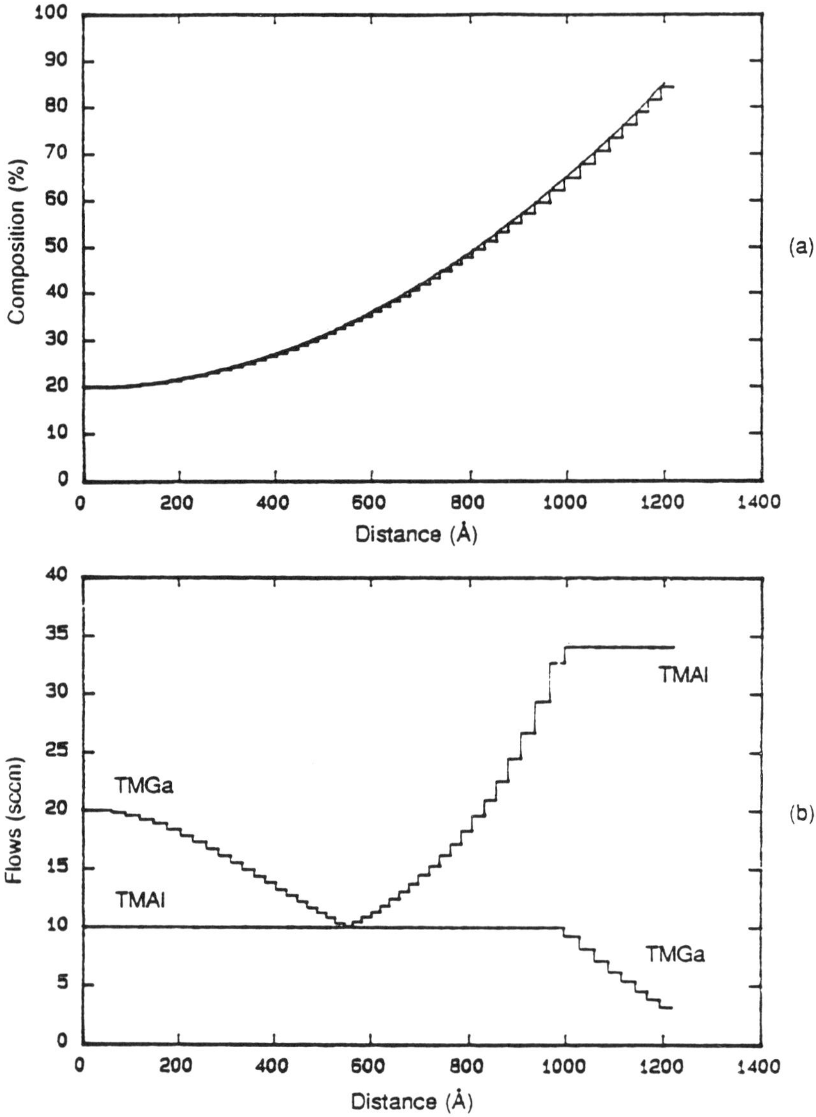

Figure 21. Grading algorithm results for a parabolically graded layer with 2-second step size and limits of 10 and 20 for TMGa and 10 and 34 for TMAl. (a) Desired (smooth curve) and algorithm-generated compositional profile. (b) Variation of TMGa and TMAl flows during layer growth.

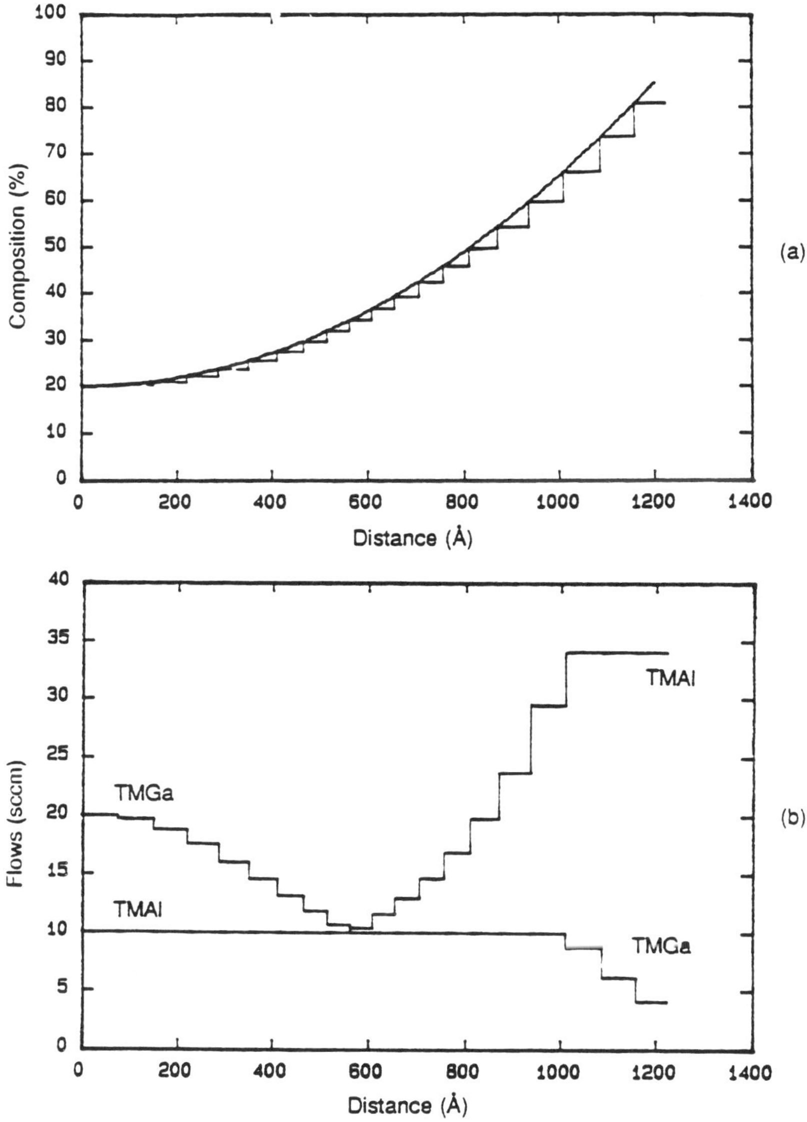

Figure 22. Grading algorithm results for a parabolically graded layer with 5-second step size and limits of 10 and 20 for TMGa and 10 and 34 for TMAl. (a) Desired (smooth curve) and algorithm-generated compositional profiles. (b) Variation of TMGa and TMAl flows during layer growth.

ments are on the order of 5 sccm as compared to approximately 2 sccm for the two-second step case. Figure 23 shows similar results for the case of a two-second time step but reactant flow limits that are entirely within the well-behaved range for the mass flow controllers. These limits are 8 and 20 sccm for TMGa and 8 and 91 sccm for TMAl. This grade is essentially identical to that shown in Fig. 21 up to a composition of approximately x=0.6. Beyond this composition the grade becomes significantly coarser owing to the larger growth rates allowed by the expanded flow range. Also note that the final two steps in the TMAl flow profile are approximately 10 and 20 sccm, changes that are too large to be accurately followed by the mass flow controller.

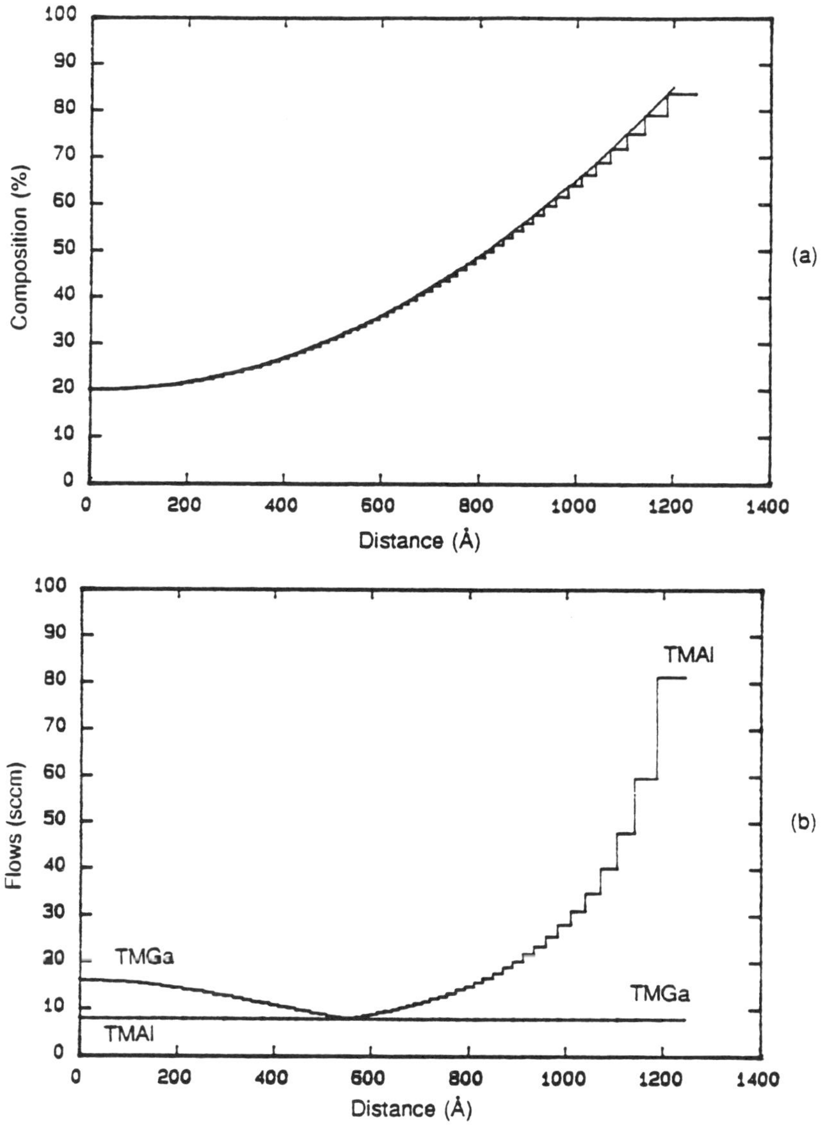

Figure 23. Grading algorithm results for a parabolically graded layer with 2-second step size and limits of 8 and 20 for TMGa and 8 and 91 for TMAl. (a) Desired (smooth curve) and algorithm-generated compositional profiles. (b) Variation of TMGa and TMAl flows during layer growth.

4. The Heterostructure Hot Electron Diode

Negative differential resistance (NDR) exists when the current-voltage characteristic of a device contains a region of negative slope, i. e., $dI/dV<0$. There are two general classes of NDR. In voltage-controlled or N-shaped NDR, the I-V characteristic contains a voltage range in which the current decreases for an increase in voltage. The voltage is thus multivalued for a certain range of current. NDR of this type is in general due to an increase in device resistance and can be brought about by a decrease in carrier mobility, carrier velocity or density of states available for conduction. In the Gunn diode,[60] electron mobility and carrier velocity are reduced above a critical electric field in some bulk direct band gap semiconductors owing to the transfer of electrons in *k*-space to indirect conduction band energy minima. In real-space transfer devices,[61,62] a similar effect is achieved when electrons traveling parallel to and on the high-mobility, narrow band gap side of a heterobarrier are heated sufficiently by an applied electric field to scatter across the barrier into a low-mobility, wide band gap region. The NDR in Esaki diodes[63] and resonant tunneling diodes[64] is due to an increase and subsequent decrease in tunneling probability in these structures with increasing voltage. In current-controlled or S-shaped NDR, the I-V characteristic contains a region in which the voltage decreases for an increase in current, and thus the current is multivalued over a certain voltage range. Devices exhibiting either type of NDR have applications as oscillators[65] and two-state logic devices.[66]

The heterostructure hot-electron diode (H^2ED)[11,67,68] is a two-terminal device based on a new switching mechanism in semiconductor heterostructures. This mechanism results in a pronounced S-shaped negative differential resistance (NDR) in the dc I-V characteristic. The results of theoretical and experimental investigations of the H^2ED, as well as data on its performance as a high-frequency oscillator, are presented in this chapter.

4.1 Theory

Consider the structure shown in Fig. 24 (a). It consists of, from left to right, a heavily doped contact region, a lightly doped narrow-gap region (GaAs:u), a lightly doped wide-gap barrier region (AlGaAs:u) and finally another heavily doped contact region (GaAs:n+). When a small positive bias is applied to the right-hand-side of the structure, electrons will drift from the left contact, across the lightly doped narrow-gap region and encounter the barrier. Since they lack sufficient energy to surmount the barrier, the only current through the device will be a tunneling current. Because this current is small, the voltage drop and electric field in the narrow-gap region are small, and thus most of the voltage drop is across the barrier region. This state is shown in Fig. 24 (b). As the bias is increased, the barrier becomes triangular and thinner and the tunneling current increases, thus increasing the field in the narrow-gap region. With increasing bias the field in this region will become large enough to heat some of the drifting electrons to the temperature necessary for thermionic emission over the barrier, resulting in an increase in current. This is an unstable point, since the increased current induces an even larger field in the narrow-gap region. Thus, the electrons will be heated even more, and the conduction mode rapidly changes from tunneling to thermionic emission. This state is shown in Fig. 24 (c). Since the tunneling mode is a relatively high resistance mode compared to thermionic emission, there will be a step change in the resistance of the device. The dc I-V characteristic of such a device should consist of a low-current tunneling regime followed by a high-current thermionic emission regime with a discontinuity at the transition point. Note that this device is not symmetrical. For a bias voltage oriented in the other direction no switch-

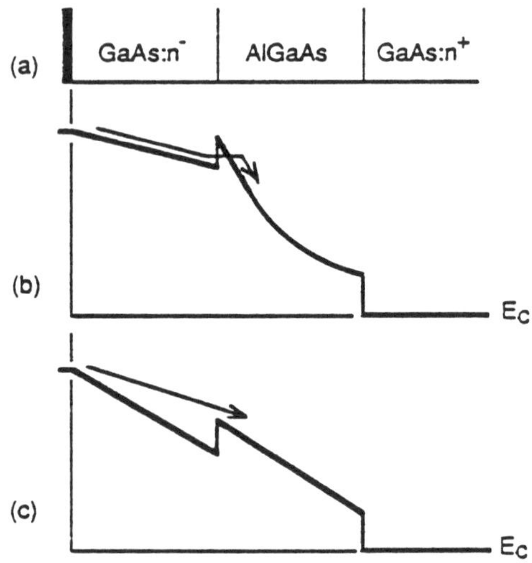

Figure 24. Heterostructure hot electron diode - (a) structure; (b) tunneling mode; (c) thermionic emission mode.

ing action is expected, since for such a bias electrons impinging on the tunneling barrier would not have first traversed a lightly doped heating region. The heating of the electron distribution in this region is essential to the switching action.

A computer simulation of this mechanism was undertaken[67] in an effort to investigate its validity. The modeled device is as described above and is shown in Fig. 25. The electric fields in the 1000 Å undoped GaAs region and 2000 Å, x=0.45 barrier region are assumed to be constant. In order for the electric fields in regions w_1 and w_2 to be different beyond a factor of the ratio of the dielectric constants, Gauss' law requires a space charge accumulation at the heterointerface. This is treated as a two-dimensional electron gas in a V-shaped potential well and is indicated as δ. This is an important feature, as it effectively lowers the barrier seen by impinging electrons by an amount E_0. The electron distribution in the GaAs heating region is described by a drifted Maxwell-Boltzmann function at electron temperature T_e, and the electron temperature is obtained by use of a power-balance equation for this region. The transmission probability through the barrier is obtained using the WKB approximation and takes into account the fact that the barrier becomes triangular and thinner with increasing applied voltage V. The details of the actual computation are published elsewhere.[67] The results of the simulation are given in Fig. 26 where the current density j and electron temperature in the GaAs region T_e are plotted against the applied voltage. As can be seen from this figure both T_e and j increase super-linearly with voltage up to approximately 5.5 volts, where there is a discontinuous increase in the current density. This discontinuity is the switching from tunneling to thermionic emission. The characteristic is not S-shaped because the simulation assumes that an ideal, zero-resistance voltage source is applied to the device. An external resistance is necessary in order for the S-shaped switch-back to be present.

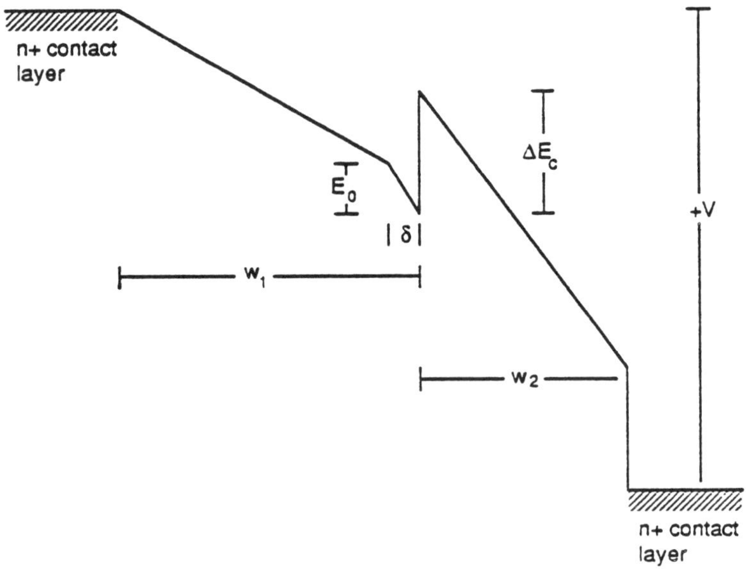

Figure 25. Conduction band edge diagram of the device structure used in computer model.

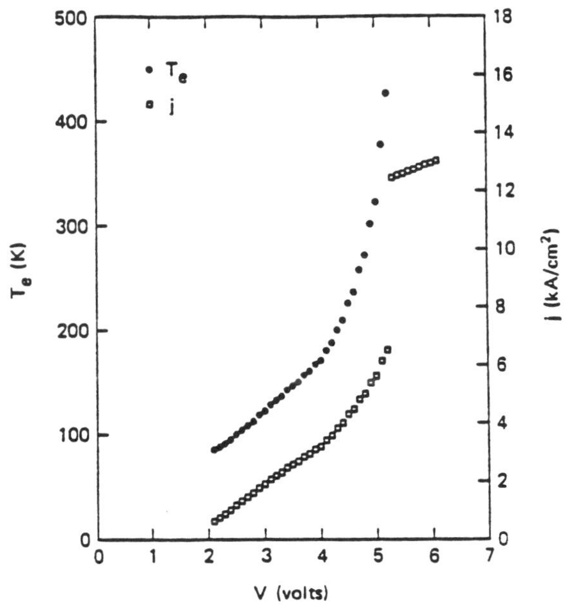

Figure 26. Calculated current density and electron temperature versus applied voltage.

4.2 Device design parameters

The device structure shown in Fig. 24 has several adjustable parameters that affect its operation: the doping and thickness of the contact layer, buffer layer, GaAs heating region and AlGaAs barrier, and the AlAs mole fraction of the AlGaAs barrier.

The structure is grown on a [100] oriented n-type conducting substrate doped with silicon at 1 to $2x10^{18}$ cm^{-3}. The first layer grown is an n-type buffer layer of approximately 0.5 μm necessary to counter the generally high substrate defect density. The doping of this layer is also 1 to $2x10^{18}$ cm^{-3} so as to make it electrically indistinguishable from the substrate. Nothing is gained from heavier doping as this layer contributes a negligible amount to the device series resistance compared to the substrate which is several mils thick. The top contact layer must be thick enough to prevent the alloyed contact from reaching the undoped GaAs region. For the shallow-diffusing gold-germanium-silver[69] contact used, 0.5 μm is an adequate thickness. The doping in this layer must be high compared to the GaAs heating region so as to contribute negligible series resistance.

The net electron concentration in the GaAs heating region should be low in order to facilitate the heating of the electron distribution, since the input power density jF, where F is the electric field, required for electron heating increases with increasing carrier concentration. This does not imply that the total impurity concentration must be low, and in fact a high degree of impurity compensation may be beneficial in order to reduce the mobility, and thus increase the resistivity, in this region. This is desirable in order to establish a larger electric field in this region for a given current flow. If the resistivity of the heating layer were extremely low, then excessive currents might be required in order to achieve a field large enough for significant electron heating. Note that due to the large electron concentration gradient between the contact and heating layers and resulting electron diffusion, the electron concentration in the undoped GaAs layer is greater than the net ionized impurity concentration there. The potential and carrier profiles of an H^2ED structure at thermal equilibrium at 77K obtained by solving Poisson's equation numerically[70] are shown in Fig. 27. All layers in this structure are 1000 Å wide. The contact layer and heating region layer dopings are 1.5×10^{18} and 10^{15} cm^{-3}, respectively, and for clarity the $Al_{0.45}Ga_{0.55}As$ barrier is assumed to have no donors or acceptors in order to prevent distortion of the figure by band-bending. The carrier concentration in the heating region varies over two orders of magnitude and at its lowest is still nearly an order of magnitude greater than the 10^{15} cm^{-3} background doping.

The doping in the AlGaAs layer affects the shape of the barrier seen by impinging electrons due to carrier transfer from the wide band gap AlGaAs to the narrow band gap GaAs and concomitant energy band-bending.[71] If the barrier doping is n-type, the effective barrier height will be reduced as the accumulation of transferred electrons at the interface causes part of the barrier to be below the Fermi level. In Fig. 28 is shown the effect of a barrier donor concentration of 5×10^{16} cm^{-3} on the structure of Fig. 27. If the barrier is p-type, the effective barrier height will be increased owing to the contact potential of the resulting p-n junction as is illustrated in Fig. 29 for the structure of Fig. 27 having a barrier acceptor concentration of 5×10^{16} cm^{-3}.

The height of the tunneling barrier is determined by the conduction band edge discontinuity ΔE_c between the heating region and the barrier which is taken to be 62% of the difference in the direct band gap Γ between the AlGaAs barrier and the GaAs.[72] The barrier band gap is direct up to an AlAs mole fraction x of approximately 0.45, and beyond that it is indirect with the X minima being the lowest in energy.[73] The possibility of participation of more than one set of minima in the tunneling process renders the exact definition of the barrier height a complex and not completely understood problem. Experimental evidence of the participation of both the X and Γ barrier valleys in tunneling processes has been observed in MOCVD-grown material.[74] Since a large tunneling barrier is desirable in the H^2ED in order to inhibit

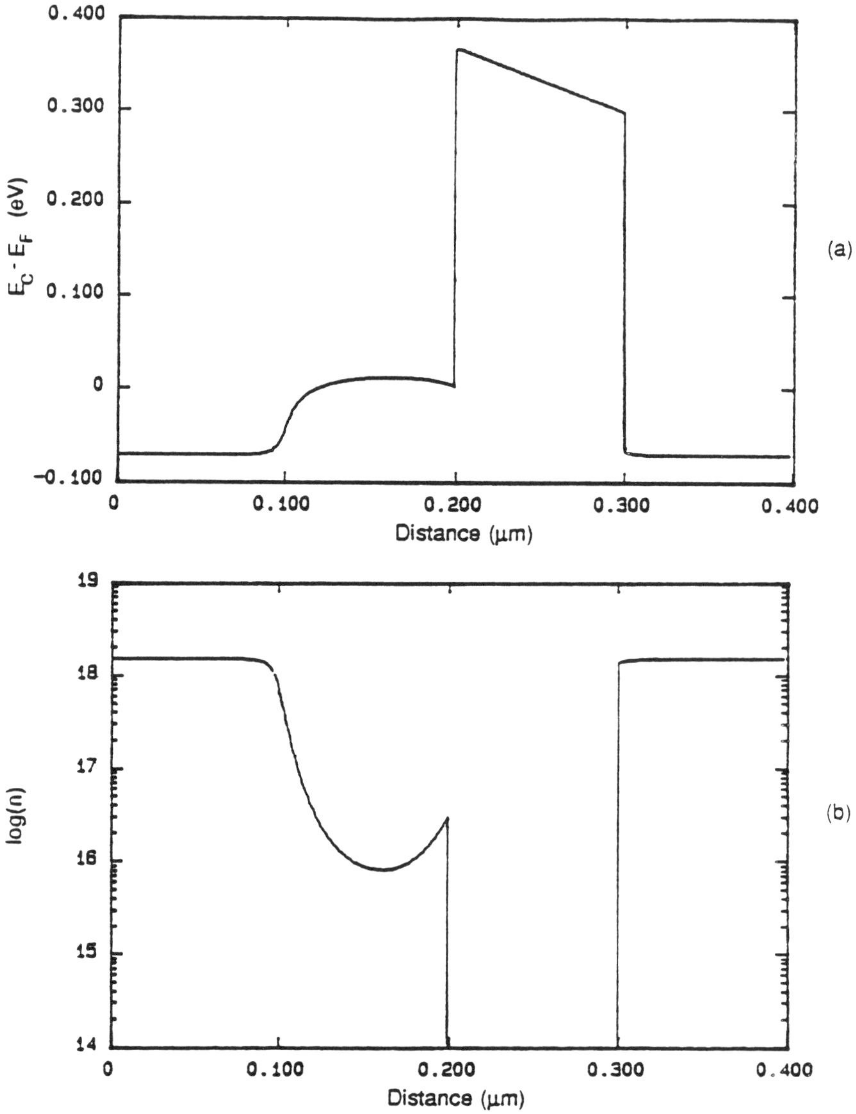

Figure 27. H^2ED equilibrium (a) conduction band edge and (b) electron profile at 77K showing effects of the heavily doped contact layer on the potential profile and carrier concentration of the undoped heating region. The donor concentration in the heating region was assumed to be 10^{15} cm^{-3}. The 45% AlGaAs barrier was assumed to have no impurities.

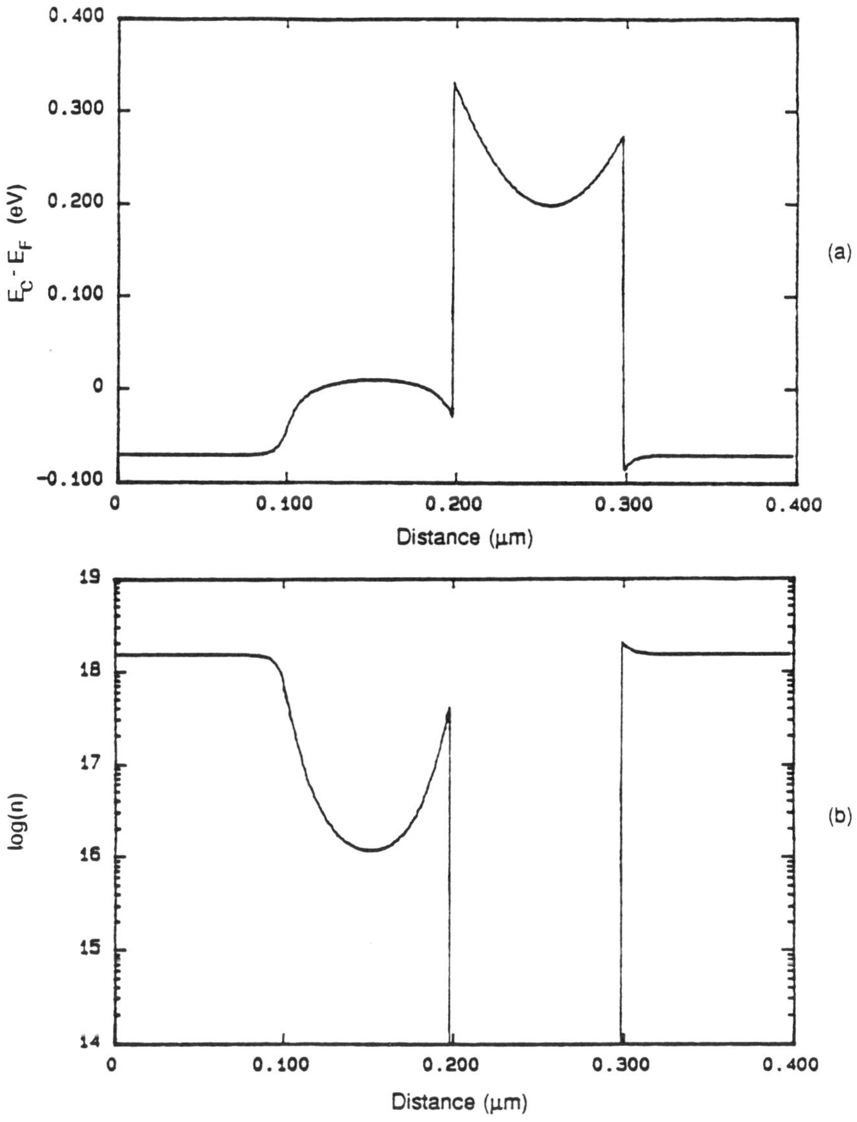

Figure 28. H^2ED equilibrium (a) conduction band edge and (b) electron profile at 77K showing effects of the barrier having a donor concentration of $5 \times 10^{16} cm^{-3}$.

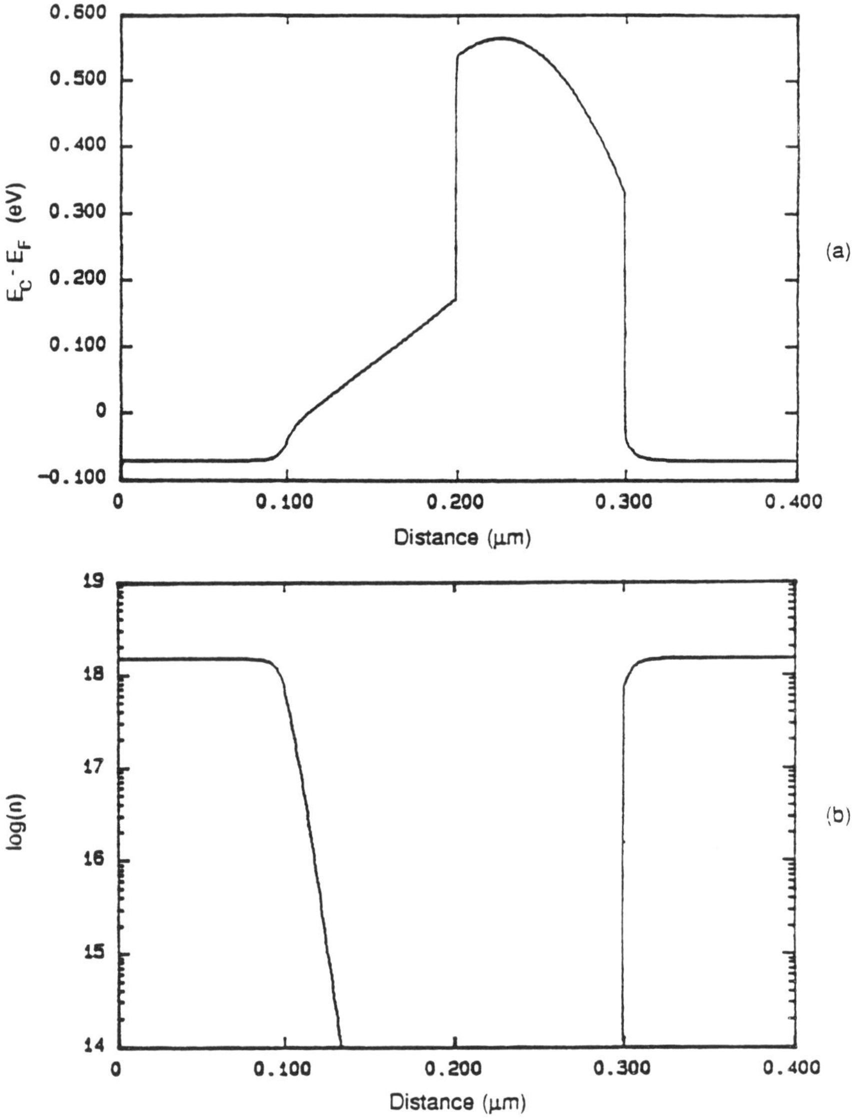

Figure 29. H^2ED equilibrium (a) conduction band edge and (b) electron profile at 77K showing effects of the barrier having an acceptor concentration of $5 \times 10^{16}\ cm^{-3}$.

thermionic emission during the low-bias, tunneling-limited portion of the dc I-V characteristic, the largest direct-gap barrier possible, corresponding to x=0.45, was chosen for preliminary investigations, giving a 77K ΔE_c of 365 meV.

The determination of the thicknesses of the heating and barrier regions is problematic in that there is an interdependence between them, the barrier height and the doping in the heating region. As yet, no sufficiently accurate quantitative model of the H^2ED has been developed to allow their determination by theoretical rather than empirical means. This interdependence may be illustrated by considering that the transition from tunneling to thermionic emission that occurs at the switching voltage can also be viewed as a transition between two different but stable electric field distributions for the same applied voltage. The field distribution in both conduction modes is affected by all the parameters just mentioned, and it is possible that for a particular combination of parameters that two such stable field distributions may not exist. Some guidelines for the empirical determination of these parameters exist. At the switching voltage a portion of the potential across the tunnel barrier is transferred to the heating region, with the remaining barrier potential being due to the IR drop across the barrier. Thus, in order for switching to occur, immediately prior to switching the barrier must have a potential drop at least equal to the sum of these two voltages. This implies that a certain minimum barrier thickness is necessary in order to ensure that an adequate voltage drop exists across the barrier at low current levels to permit switching. If such a voltage requires high current, then the IR drop across the heating region might be large enough to allow both thermionic emission and tunneling to occur, thus precluding the H^2ED switching mechanism. A thick barrier has a voltage range over which the current is extremely small (<1 μA), followed by a rapid increase in current with increasing voltage. Since in the low-current range essentially all of the voltage drop is across the barrier, this range can be considered to be a barrier potential accumulation range. For an x=0.45 barrier at 77K, the effective barrier thickness at diode turn-on is approximately 300 Å. To ensure a low-current regime in the I-V characteristic, the barrier should be at least that wide. There is no maximum barrier thickness, although the higher switching voltages associated with wider barriers may lead to excessive power dissipation and thermal damage to the device. Also, since the switching speed is expected to be transit-time limited, the device should be as short as possible. Once a barrier height and thickness have been chosen, an appropriate heating region thickness can be determined empirically.

4.3 Experiment

Experimental structures were fabricated in order to verify the switching mechanism. Device wafers were grown by atmospheric pressure metalorganic chemical vapor deposition (MOCVD) on [100] oriented conducting substrates (Si, $2x10^{18}$) using trimethylgallium, trimethylaluminum, arsine and hydrogen selenide. The structure consisted of a GaAs:n+ buffer, followed by 2000 Å of $Al_{0.45}Ga_{0.55}As$:u and finally 2000 Å of GaAs:u and is illustrated in Fig. 30. The growth rates were 6 and 11 Å/s and the V-III ratios were 40 and 22 to 1 for the GaAs and AlGaAs, respectively. Gold-germanium-silver contacts were defined by liftoff metallization, followed by mesa definition and etch.

A typical device dc I-V characteristic is shown in Fig. 31. In order to avoid lattice heating effects, pulsed I-V measurements were performed with a pulse width of 300 ns and a repetition rate of 100 Hz. As can be seen, the anticipated S-shaped NDR is present. This measurement was performed at a temperature of 16K. The switching was observed at 77K, but was enhanced at lower temperatures. It should be noted that for the opposite bias, in which electrons are incident on the barrier without first traversing a lightly doped heating region no switching was observed, which is consistent with the proposed model.

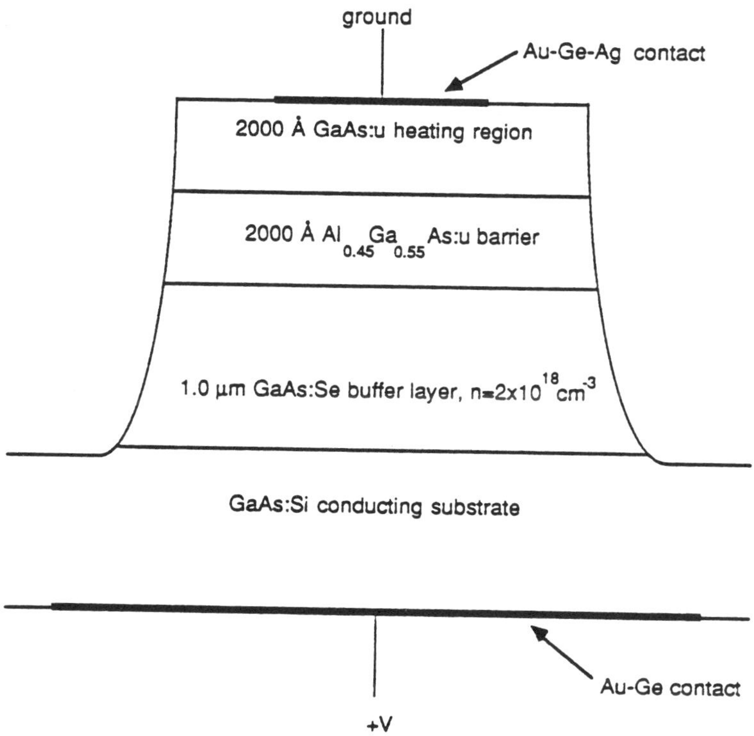

Figure 30. Schematic of single-barrier H^2ED.

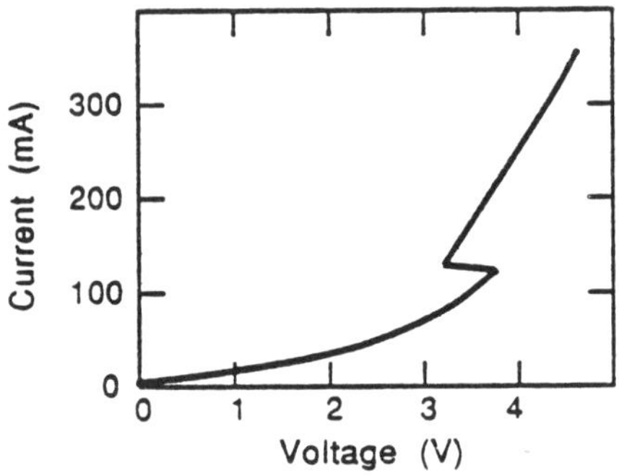

Figure 31. Experimental H^2ED I-V characteristic. The measurement was performed at 16K using 300 ns voltage pulses at a 100 Hz repetition rate.

In an effort to reduce the magnitude of the current prior to switching, a multibarrier design, shown in Fig. 32 (a), was implemented. Conduction band edge diagrams of this structure for both conduction modes are shown in Figs. 32 (b) and (c). Such a structure should display the same switching behavior as a single barrier device, but with a higher switching voltage owing to a greater fraction of the applied bias being dropped across the wider effective barrier. A typical 77K I-V characteristic of such a device is shown in Fig. 33. The current is extremely small until the bias is increased to within a few volts of the switching point, which is an indication of the effectiveness of the multiple tunnel barrier approach. This effectiveness may be due to electrons being forced to undergo three separate tunneling events in order to traverse the device, thus reducing the total tunnel current.[67] For the reverse bias no switching was ever observed.

In all such devices tested, increasing the current to a level much higher than the switching current in either bias direction caused light emission at the wavelength corresponding to the wavelength of bulk GaAs. Since the structure is entirely n-type, the holes necessary for this recombination radiation must originate in impact ionization. This is to be expected, since the multibarrier H^{2}ED is structurally similar to a superlattice avalanche photodiode.[75] If it is assumed that the 16-volt switching voltage is dropped uniformly across the portion of the device between the n+ layers (0.57 μm), an average electric field of 281 kV/cm is obtained which is more than sufficient for impact ionization to occur in the wells.[75] By observing the optical emission spectra of devices having low-composition AlGaAs in the buffer, wells or cap layer, it was determined that the recombination was occurring in the contact region into which the generated holes were swept by the electric field. No recombination radiation was observed to occur in the wells. In an attempt to reduce the impact ionization, the devices were modified such that the second and third wells were reduced in thickness from 900 Å to 450 Å and the barriers were made larger to keep the same overall

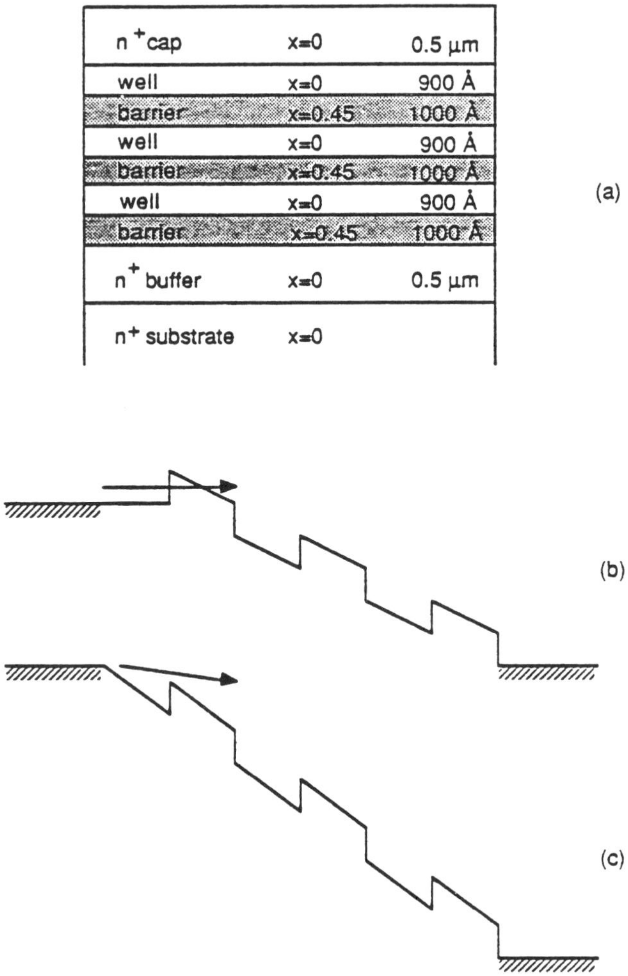

Figure 32. (a) Multi-barrier H^2ED structure; (b) Conduction band edge in preswitching tunneling mode; (c) Conduction band edge in post-switching thermionic emission mode.

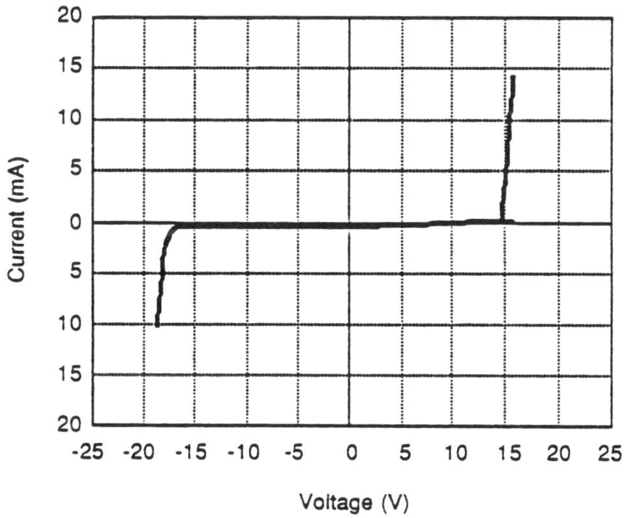

Figure 33. Current-voltage characteristic at 77K for H^2ED structure of Fig. 32 showing S-shaped switching in forward bias and a smooth turn-on in reverse bias. Device diameter is 6 mils.

thickness. The reduction in well thickness significantly reduced the impact ionization coefficient,[75] and the amount of emitted light was significantly less than for the previous structure. The switching mechanism is believed to be independent of the impact ionization because light emission occurs identically for both bias directions but switching only occurs for forward bias, and because the impact ionization can be reduced through device design modifications without reducing the switching effect.

4.4 Crystal growth considerations

Successful implementation of any device structure by MOCVD requires careful determination of the optimum growth parameters. The versatility of MOCVD is due in part to the fact that material characteristics can be tailored over a wide range to specific applications by adjustment of growth parameters. The critical layers in the H^2ED are the undoped GaAs electron heating region and the AlGaAs tunnel barrier, and considerations related to their growth will now be discussed.

The nominally undoped GaAs electron heating layer must have a low net carrier concentration in order to allow heating of the electron distribution in this region at low input power densities. Without a quantitative analytic model it is impossible to quantify this, but certainly the carrier concentration must be less than that in the heavily doped contact layers, about 10^{18} cm^{-3}. Electrons should be the majority carriers in this region so as to prevent the formation of a p-n junction and associated potential barrier at the interface between the heating region and contact layer. Because of the diffusion of electrons from the contact layer into to the heating region, the heating region can have a p-type residual doping concentration up to the concentration of electrons originating in the contact layer, but the net carrier type in the region will still be n-type. Note that the net electron concentration is the primary concern, with the compensation ratio and mobility being secondary concerns as discussed in Section 4.2. This is due to the fact that impurity scattering is an elastic process[76] and thus does not directly affect carrier heating. The carrier concentration affects carrier heating, however, by determining the input power necessary to achieve a given carrier temperature, with a lower impurity concentration requiring less input power. This affords flexibility in the selection of growth temperature since undoped GaAs can be kept n-type with a carrier concentration in the 10^{15} cm^{-3} range over much of the typical growth temperature range of 600-800 °C through adjustment of the arsine to trimethylgallium (V-III) ratio. If high mobility and low total impurity concentration were critical concerns, the growth temperature would have been restricted to the lower portion of this range with the highest mobility occurring in the 600-650 °C range, depending on the purity of the sources used.[38]

The $Al_{0.45}Ga_{0.55}As$ tunnel barrier layer, as discussed earlier, is of critical importance to the operation of the H^2ED. The characteristics of AlGaAs grown by MOCVD vary greatly with growth temperature. There are a few important general trends to be considered. First, the concentration of deep levels, or traps, in AlGaAs increases with decreasing growth temperature.[13,77] Such traps can act as non-radiative minority carrier recombination centers[77,78] and for this reason optical devices such as heterostructure lasers are usually grown at the higher end of the temperature range mentioned above. Second, the p-type background concentration of AlGaAs increases with increasing growth temperature. This is illustrated in Fig. 34. Such behavior has been attributed to the incorporation of carbon into the growing layer from decomposed organic radicals due to the strong aluminum-carbon bond.[13] A similar carbon incorporation trend has been observed in GaAs but to a much lesser extent owing to the weaker gallium-carbon bond.[79] Below a certain growth temperature insufficient shallow acceptors are incorporated to counter the number of deep levels, and the material becomes electrically compensated and extremely resistive. For the $Al_{0.45}Ga_{0.55}As$ shown in Fig. 34 this temperature is approximately 720 °C.

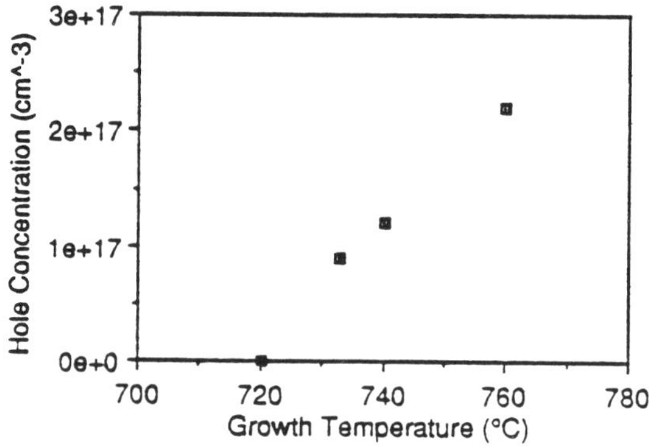

Figure 34. Hole concentration versus growth temperature for undoped AlGaAs, x=0.45.

Growth condition optimization was done first for growth temperature which, as described above, affected primarily the $Al_{0.45}Ga_{0.55}As$ tunnel barrier. Identical device structures were grown over a temperature range of 700-750° C with a temperature interval of 10° C. It was found that the growth temperature at which the devices consistently displayed the switching behavior was 720° C. Very few devices from wafers grown at other temperatures displayed the switching at all, and those that did displayed the switching to a smaller degree. Devices grown at lower temperatures tended to undergo catastrophic breakdown and to become short circuits before the switching threshold was achieved. This temperature dependence was checked several times for slightly difference device structures, and the optimum growth temperature was consistently in the 720-725° C range. This behavior may be explained in light of the data in Fig. 34. The optimum growth temperature range coincides with the growth temperature at which the $Al_{0.45}Ga_{0.55}As$ becomes depleted of free carriers. As discussed in Sec. 4.2, this is the desired situation so as to avoid distortion of the barrier by carrier transfer effects. The failure of devices grown at lower temperatures may be attributed to excessive power dissipation in, and subsequent thermal damage to, the highly resistive barrier layer.

Optimization of the electron concentration in the GaAs heating region was undertaken next. Initial H^2ED wafers grown had an electron concentration of $2x10^{16}$ cm^{-3} in the heating region, based on measurements made on bulk GaAs layers grown under similar conditions. Devices fabricated from such wafers exhibited the switching effect. Devices in which the heating region was deliberately doped to $n=5x10^{17}$ cm^{-3} did not exhibit the switching, and the resulting I-V characteristics displayed identical smooth diode turn-ons for both forward and reverse biases. Further device wafers grown with heating region carrier concentrations of $n=2x10^{15}$ and 0 cm^{-3}, achieved by altering the arsine to TMGa (V-III) ratio, yielded devices that exhibited the switching nearly identically. It was thus concluded that the H^2ED switching effect is insensitive to the exact carrier concentration in the lightly doped heating region provided it is below some threshold level, with the allowable range including concentrations less than or equal to $2x10^{16}$ cm^{-3}. This conclusion is in agreement with computer simulations similar to those described in Section 4.1.[80]

The absence of switching in the reverse bias direction offers strong experimental support for the validity of the proposed switching mechanism. However, there was another effect that could have been responsible for this asymmetry in the forward and reverse I-V characteristics. The quartz growth chamber and silicon carbide-coated carbon susceptor on the MOCVD reactor used in this work are removed and cleaned after each growth, thus exposing them to air and water. The cleaning procedure involves etching deposited material, mostly arsenic, with a hydrofluoric:nitric acid mixture (2:9), rinsing with de-ionized water and methanol, and drying in a dry nitrogen gas stream. The susceptor is cleaned in a similar manner. After a substrate is loaded into the cleaned chamber, the chamber is evacuated and then purged with palladium-purified hydrogen. Even after this procedure there still may be enough residual water and oxygen present in the chamber to adversely affect the growth of AlGaAs. The effects of such residual contamination of the growth environment can be reduced by use of a low Al composition AlGaAs buffer layer.[81] The highly reactive TMAl in such a buffer layer reactant stream reacts with and removes the residual water and oxygen from the growth environment, thus leaving a contaminant-free environment for the growth of subsequent AlGaAs layers. In the process, contaminant-related traps are incorporated in the AlGaAs buffer layer rather than in critical device layers. If no getter layer is used then the initial AlGaAs device layers grown will inadvertently serve the same purpose, possibly with adverse effects on device performance. Initial investigations were done on H^2ED structures such as those shown in Figs. 30 and 32 in which the critical interface between the first heating layer and barrier was grown only after a substantial thickness of AlGaAs was grown.

Thus, the AlGaAs near this interface benefited from the gettering effects of the previously grown AlGaAs. The first AlGaAs grown, the barrier on the heavily doped GaAs buffer layer, would have suffered most from gettering residual contaminants. It was possible that the H^2ED failed to exhibit switching in reverse bias because this barrier was of poor quality, rather than for the reasons put forth in the proposed model of the switching mechanism. In order to resolve this issue, an inversion of the multi-barrier H^2ED structure of Fig. 30 was grown in which the undoped GaAs heating layer was the first layer grown on the heavily doped buffer. In this orientation the critical switching interface is the first AlGaAs/GaAs interface grown, and the switching is expected to occur with the positive bias on the top contact. Devices fabricated from this wafer exhibited the switching only in the expected bias, thus eliminating the possibility that the switching asymmetry in the forward and reverse bias I-V characteristics was an artifact of the layer growth order.

4.5 The NERFET

It is interesting to note the similarities between the H^2ED and the negative resistance field effect transistor (NERFET).[62,82,83] The NERFET is a three-terminal device exhibiting N-shaped NDR. The device wafer is grown on an n-type substrate and consists of a GaAs:n+ buffer, a 2000 Å $Al_{0.45}Ga_{0.55}As$ tunnel barrier and 2000 Å of GaAs:u. Device fabrication consists of depositing a backside ohmic contact, referred to as the collector, followed by Au-Ge-Ag ohmic source and drain contacts on the wafer surface. The source and drain contacts are carefully alloyed into, but not beyond, the barrier so that they contact the top barrier interface without shorting through to the conducting substrate. The devices are mesa isolated. Diagrams of an as-grown wafer and a fabricated device are shown in Fig. 35. In operation, the drain contact is tied to ground, and the source and substrate contacts are connected to separate voltage sources. Application of a positive substrate voltage of a few volts induces a two-dimensional electron gas on the surface side of the barrier, thus providing a conducting channel between source and drain. As a positive increasing voltage is applied to the source, the source-drain current increases and the carrier distribution in the channel is heated by the applied electric field. When the distribution is heated to the point that thermionic emission over the barrier is probable, current is diverted over the barrier to be collected by the substrate contact. This gives rise to NDR in the source-drain I-V characteristic which is exemplified in Fig. 36. This mechanism is essentially the same as the real-space transfer proposed and demonstrated by Hess et al., but with the addition of the collector terminal to enhance the NDR.

Although the details of the NDR mechanisms in the H^2ED and NERFET differ, they share a fundamental aspect. In both devices the NDR is triggered by the onset of thermionic emission over a barrier caused by the heating of the carrier distribution by an applied electric field. Because of this similarity, material considerations for these devices are remarkably similar. The top undoped GaAs layer in the NERFET is functionally analogous to the undoped GaAs heating region in the H^2ED in that both are regions where electron heating occurs, and as such both are tolerant of low mobility and high total impurity concentration. The tunnel barrier in both devices is extremely critical. In the NERFET, a source-collector diode that is leaky or that has soft turn-ons is an indication that the NERFET will not display NDR. It was found that growth conditions identical to those determined for the H^2ED were necessary in order to produce working devices, with 720° C being the optimum growth temperature, reflecting the sensitivity of the device to the quality of the tunnel barrier.

4.6 Comparison to other models

The multibarrier H^2ED is similar to a device structure reported by Belyantsev et al.[84] that exhibits similar NDR behavior, but the switching mechanism proposed for

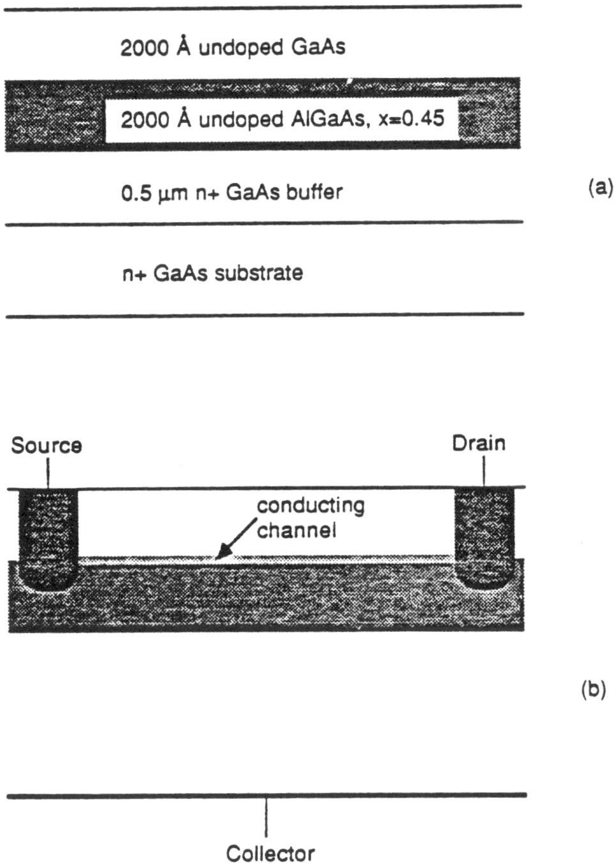

Figure 35. NERFET (a) as-grown wafer and (b) device structure. Typical source-drain spacing is 1µm.

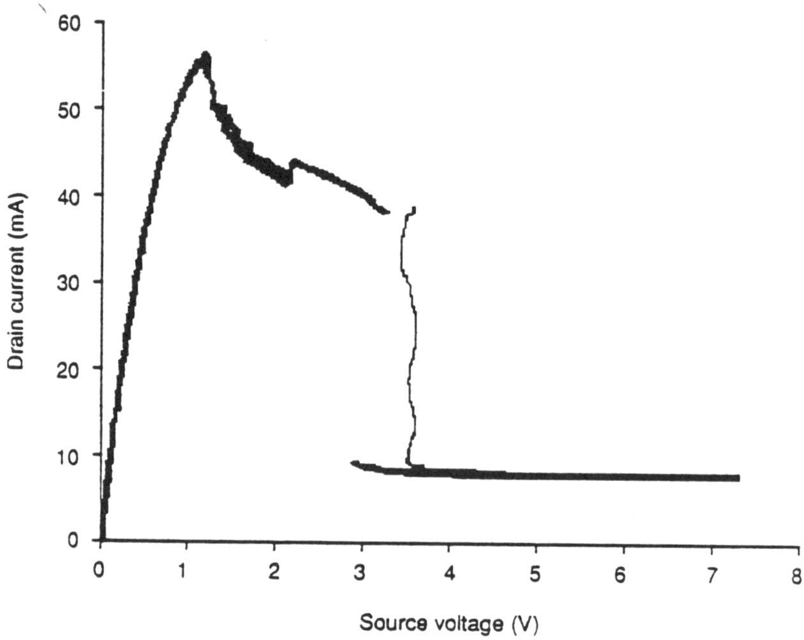

Figure 36. NERFET I-V characteristic at 77K for a device with a source-drain spacing of 1.5 microns, a 300-micron drain and a collector voltage of 4.4 V.

this device is different than that proposed by Hess et al.[11] In this alternate model,[84-87] each of the three sections of the S-shaped I-V characteristic corresponds to a different regime of electron heating in the GaAs wells. In order that electrons be present in the wells for this effect to occur, it is required that the wells be doped with a donor concentration in the 10^{16} to 10^{17} cm^{-3} range so that hot electrons entering the well can be rapidly thermalized through collisions with other electrons and contribute their excess energy toward the heating of the distribution. The heating of the electrons in the wells occurs when an electron in a barrier layer gains an energy E_{drift} from the applied electric field and enters a well, acquiring an additional energy ΔE_{c1} from the conduction band discontinuity, thus contributing a total energy of $E_{drift}+\Delta E_{c1}$ to the total electron energy in the well. An electron must then leave the well by thermionic emission, taking with it an energy ΔE_{c1}. Thus the net energy of the electron distribution in the well is increased by an amount $\Delta\varepsilon=E_{drift}$. As the current increases, the number of such incoming hot electrons heating the electron distribution in the well increases, thus making the electron temperature in the well a function of the current. The electron temperature in the well can be found by the energy balance equation

$$j\,\Delta\varepsilon/e = (3nL/2)\,k\,(T_e-T_o)\,/\,\tau_\varepsilon \qquad (4.1)$$

where j is the current density, e is the electronic charge, n is the density of electrons in the well, L is the well width, k is Boltzmann's constant, T_e is the electron temperature, T_o is the lattice temperature which is assumed to be constant, and τ_ε is the electron energy relaxation time. The left-hand side of this equation represents the rate of energy input into the well by hot electrons, and the right-hand side is the rate of energy loss to the lattice by the electrons in the well. Tunneling is assumed to be insignificant due to the thickness of the barriers (~1000 Å), leaving thermionic emission as the only escape mechanism for electrons in the well. The current density is then described by a simple expression for thermionic emission with T_e substituted for T_0

$$j = e\, v_R\, n \exp(e\, \varphi_b / k T_e) \qquad (4.2)$$

where v_R is the rate of thermionic emission and φ_b is the barrier seen by electrons in the well. This expression ignores the thermionic emission current from the barrier to the well, which is taken to be negligible compared to the forward current. There is no explicit dependence on the applied voltage since φ_b is an abrupt heterobarrier and hence not affected. The current density is a function only of T_e, which in turn is a function of the current density and $\Delta\varepsilon$. Combining Eqs. (4.1) and (4.2) to eliminate T_e yields

$$\Delta E = \alpha\, (-\Phi_b / \ln(I) - 1) / I \qquad (4.3)$$

where ΔE is the net energy $\Delta\varepsilon$ contributed to the electron distribution in the well by each incoming electron in units of $k T_0$, $\alpha = L/(\tau_\varepsilon v_R)$, Φ_b is the exit barrier height in units of $k T_0$, and $I = j/(e\, n v_R)$ is the dimensionless current. In this model, the applied voltage U per period of the structure is assumed to be dropped entirely across the barrier, so the energy $\Delta\varepsilon$ acquired by each electron in the wide gap region is equal to eU. Thus, Eq. (4.3) describes an I-V characteristic. Such an I-V characteristic is shown in Fig.37 for a structure with 600 Å wells and barrier heights of $10kT_0$, and is S-shaped. A similar analytic result is also derived for the case of a similar structure with graded instead of abrupt barriers.

This analytic model has two major flaws. The definition given for $\Delta\varepsilon$ is only valid for $T_e = T_0$. Specifically, an electron leaving the well takes with it an energy $E_{c1}+E_{hot}$ where E_{c1} is the barrier height and E_{hot} is the electron's energy in excess of the

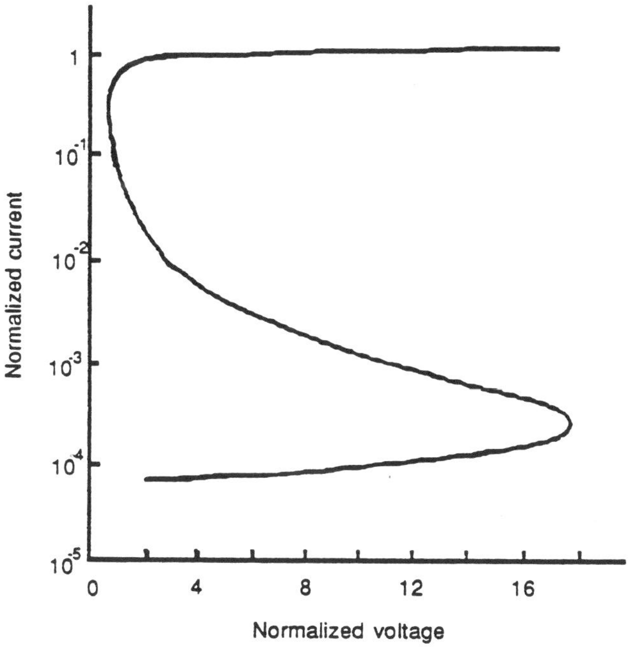

Figure 37. Plot of Eq. 4.3 showing S-shaped NDR following from model of Belyantsev et al. for a device structure with 600 Å wells and barriers of 10kT. (After reference 83)

barrier height. The excess energy E_{hot} is dependent on the distribution and increases with T_e.[62] Since $\Delta\varepsilon$ is defined as the net energy contribution to the distribution in the well by each incoming electron, its defining relation becomes $\Delta\varepsilon = E_{drift} + E_{hot}$. With $E_{drift} = eU$ as mentioned above, $U = (\Delta\varepsilon + E_{hot})/e$ is the equation necessary to generate I-V relationships from Eq. (4.3). The addition of the E_{hot} term acts to counter the NDR shown in Fig. 37. Finally, tunneling current is ignored. Even though the barriers are thick, effective barrier thinning by the applied field will result in significant tunneling current at the high electric fields near the switching voltage.

The proposed physical mechanism responsible for the S-shaped NDR is not made clear. It is stated that each of the branches of the S-shaped I-V characteristic corresponds to a different T_e regime in the wells. During the low current portion of the I-V characteristic, T_e is of order T_0. As the NDR section of the I-V characteristic is approached, T_e is of order $2T_0$ and there is a sudden increase in thermionic emission owing to the hotter electron distribution, resulting in the NDR. No explanation is offered for this sudden increase, and it would appear to be nonphysical in that the electron distribution is always smoothly varying and no mechanism is present for an abrupt change in T_e. The final branch of the I-V curve is a high T_e regime in which nearly all electrons in the well possess adequate energy to surmount the barrier, resulting in a nearly ohmic continuation of the I-V characteristic.

A comparison of the H^2ED model of Hess et al. discussed in Sec. 4.1 and thermionic emission (TE) model just described reveals a few similarities and several major differences. In both models, the switching is from a high resistance, low current initial state to a lower resistance, high current thermionic emission state. Also, the switching in both models is due to the onset of thermionic emission over a barrier after the heating of the electron distribution. However, in the TE model, the heating occurs when electrons acquire energy while drifting across an undoped barrier, while in the H^2ED the heating occurs when electrons participating in a tunneling current acquire energy while drifting across an undoped GaAs heating region. It can be argued that the heating in the H^2ED could be due to the same mechanism as in the TE, but such a contention fails in light of the operation of the single barrier H^2ED in which there is no barrier prior to the undoped GaAs region. In addition, the fact that the switching in the H^2ED occurs only in one bias direction is strong evidence of the validity of the proposed switching mechanism. The TE model predicts identical switching for both bias directions, and no mention of the presence or absence of such symmetry is made in the published results of Belyantsev et al. The TE model requires that the GaAs well regions be doped to an unspecified minimum concentration, with doping levels in the 10^{16} to 10^{17} cm^{-3} range mentioned, while experimental results with the H^2ED have shown NDR at GaAs doping levels of $2x10^{16}$, $2x10^{15}$ and ~0 cm^{-3}. Finally, the TE model offers no physical mechanism for the switching transition, while the H^2ED model proposes a positive feedback mechanism in which the conduction mode changes completely from tunneling to thermionic emission at a threshold bias voltage.

4.7 Oscillation results

The pronounced NDR of the H^2ED suggests its application as an oscillator. During initial dc characterization oscillation at frequencies greater than those measurable with a laboratory oscilloscope were observed when devices were biased near their switch point with a curve tracer. In the proposed model of the switching mechanism, the time necessary for the H^2ED to switch on is the transit time of an electron across the device. Thus, assuming a saturation velocity of $2x10^7$ cm/s, a device such as that in Fig. 32 with a length of 0.57 μm would have a theoretical upper frequency limit of approximately 175 GHz. The actual frequency limit may be lower owing to parasitic effects such as packaging capacitance and inductance.

In order to characterize the H^2ED as a free-running oscillator, a test system was set up[89] as shown in Fig. 38. In this system, the device is biased near the switch point

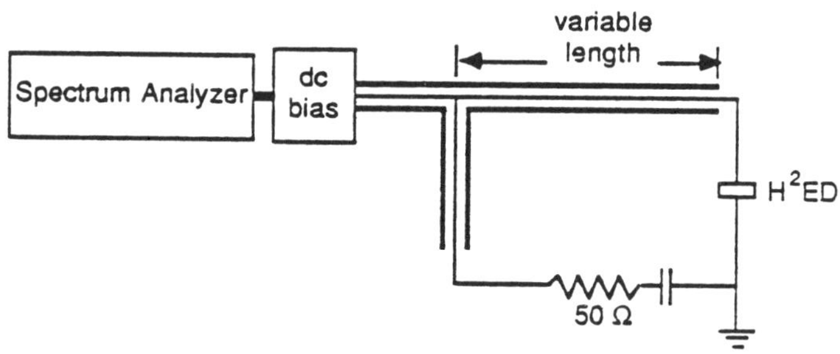

Figure 38. Block diagram of system used for free-running oscillator characterization.

in a cavity defined by a 50 ohm coaxial cable of any desired length. The device structure used in this experiment was similar to that of Fig. 32 except that the top contact layer was $Al_{0.2}Ga_{0.8}As$ instead of GaAs, and two of these structures were grown in series on the same wafer. This stacking was done to reduce the capacitance of the completed 3 mil diameter devices. The devices were mounted on an SMA connector and contacted with a tungsten whisker. Measurements were made at 77K. Figure 39 is a typical frequency spectrum measured in this way and clearly shows cavity mode structure and oscillation to at least 5 GHz.

Further characterization was done[89] with a Hewlett-Packard 8510 network analyzer to investigate H^2ED performance as an amplifier. A multibarrier structure similar to that in Fig. 32 was fabricated into 3 mil diameter devices and mounted in a 50 ohm strip line. Reflection gain was measured at 77K with the device biased near its switch point. An example of a gain spectrum obtained is given in Fig. 40. Broad band amplification was obtained with a peak gain of 3.75 dB at 15.7 GHz and a unity gain cutoff of 17.2 GHz. An abrupt roll-off at higher frequencies was observed and was probably due to the 18 GHz upper frequency limit of the test fixture.

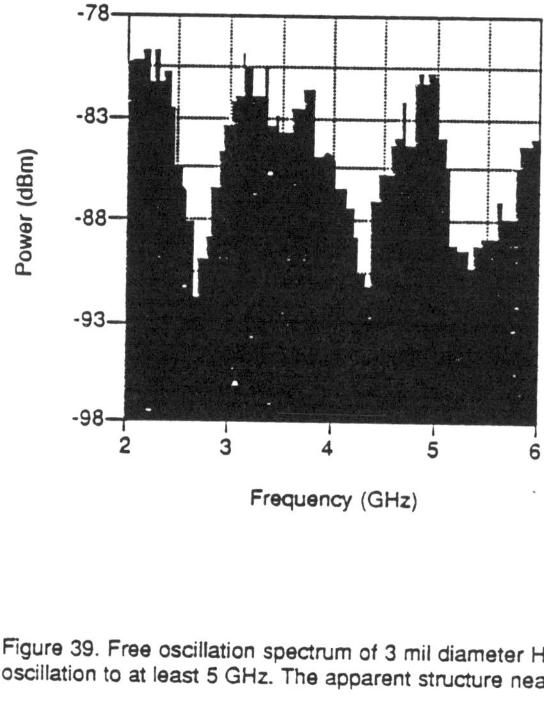

Figure 39. Free oscillation spectrum of 3 mil diameter H²ED showing oscillation to at least 5 GHz. The apparent structure near 6 GHz is noise.

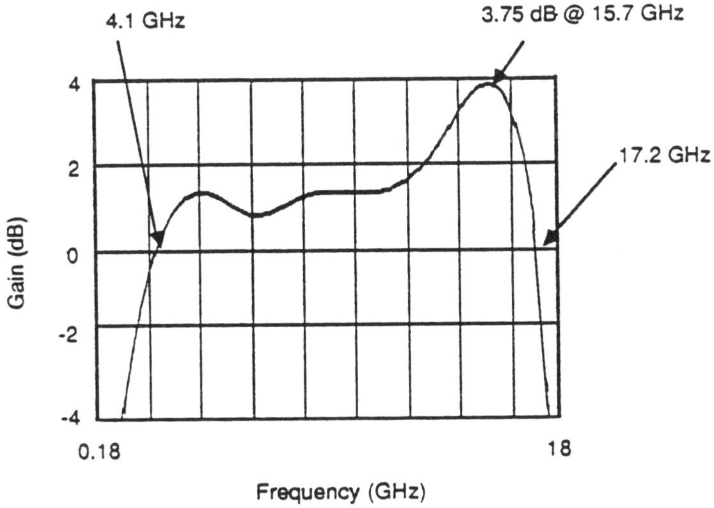

Figure 40. S_{11} reflection gain for a 3 mil diameter H^2ED similar to that shown in Fig. 32.

References

1. H. M. Manasevit, Appl. Phys. Lett. **12**, 156 (1968).

2. J. J. Coleman and P. D. Dapkus, **Gallium Arsenide Technology**, (Howard W. Sams & Co., Inc., Indianapolis, 1985), p. 79.

3. L. M. Miller and J. J. Coleman, CRC Critical Reviews, in press.

4. T. K. Higman, S. J. Manion, I. C. Kizilyalli, M. A. Emanuel, K. Hess and J. J. Coleman, Phys. Rev. B **36**, 9381 (1987).

5. J. J. Coleman, P. D. Dapkus and J. J. J. Yang, Elect. Lett. **17**, 606 (1981).

6. R. Bhat, W. K. Chen, A. Kastalsky, M. A. Koza and P. S. Davisson, Appl. Phys. Lett. **47**, 1344 (1985).

7. S. J. Jeng, C. M. Wayman, G. Costrini and J. J. Coleman, J. Cryst. Growth **73**, 425 (1986).

8. R. D. Dupuis and P. D. Dapkus, Appl. Phys. Lett. **31**, 466 (1977).

9. L. J. Mawst, M. E. Givens, M. A. Emanuel, C. A. Zmudzinski and J. J. Coleman, J. Appl. Phys. **60**, 2633 (1986).

10. L. J. Mawst, M. E. Givens, C. A. Zmudzinski, M. A. Emanuel and J. J. Coleman, IEEE J. Quantum Electron. **QE-23**, 698 (1987).

11. K. Hess, T. K. Higman, M. A. Emanuel and J. J. Coleman, J. Appl. Phys. **60**, 3775 (1986).

12. K. L. Hess, D. L. Kasemset and P. D. Dapkus, J. Electron. Mater. **13**, 779 (1984).

13. T. F. Kuech, D. J. Wolford, E. Veuhoff, V. Deline, P. M. Mooney, R. Potemski and J. Bradley, J. Appl. Phys. **62**, 632 (1987).

14. T. E. Schlesinger and T. Kuech, Appl. Phys. Lett. **49**, 519 (1986).

15. M. Mizuta, T. Iwamoto, F. Moriyama, S. Kawata and H. Kukimoto, J. Cryst. Growth **68**, 142 (1984).

16. R. Bhat, M. A. Koza, C. C. Chang and S. A. Schwarz, J. Cryst. Growth **77**, 7(1986).

17. R. H. Moss, J. Cryst. Growth **68**, 78 (1984).

18. P. D. Dapkus, H. M. Manasevit, K. L. Hess, T. S. Low and G. E. Stillman, J. Cryst. Growth **55**, 10 (1981).

19. H. Terao and H. Sunakawa, J. Cryst. Growth **68**, 157 (1984).

20. M. S. Kim, S. Min and J. S. Chun, J. Cryst. Growth **74**, 21 (1986).

21. D. W. Vook, S. Reynolds and J. F. Gibbons, Appl. Phys. Lett. **50**, 1386 (1987).

22. R. Bhat, M. A. Koza and B. J. Skromme, Appl. Phys. Lett. **50**, 1194 (1987).

23. R. M. Lum, J. K. Klingert and M. G. Lamont, Appl. Phys. Lett. **50**, 284 (1987).

24. R. J. Field and S. K. Ghandhi, J. Cryst. Growth **74**, 551 (1986).

25. H. Asai and H. Sugiura, Jap. J. Appl. Phys. **24**, L815 (1985).

26. A. J. SpringThorpe, F. D. King and A. Becke, J. Electron. Mater. **4**, 101 (1975).

27. E. Veuhoff, T. F. Kuech and B. S. Beyerson, J. Electrochem. Soc. **132**, 1958 (1985).

28. J. D. Parsons and F. G. Krajenbrink, J. Cryst. Growth **68**, 60 (1984).

29. J. S. Roberts, N. J. Mason and M. Robinson, J. Cryst. Growth **68**, 422 (1984).

30. H. Ohno, E. Ikeda and H. Hasegawa, J. Cryst. Growth **68**, 15 (1984).

31. R. W. Glew, J. Cryst. Growth **68**, 44 (1984).

32. C. R. Lewis, W. T. Dietze and M. J. Ludowise, J. Electron. Mat. **12**, 507 (1983).

33. M. L. Timmons, P. K. Chiang and S. V. Hattangady, J. Cryst. Growth **77**, 37 (1986).

34. J. D. Parsons, L. S. Lichtmann and F. G. Krajenbrink, J. Cryst. Growth **77**, 32 (1986).

35. G. Costrini and J. J. Coleman, J. Appl. Phys. **57**, 2249 (1985).

36. C. A. Wang, S. H. Groves, S. C. Palmateer, D. W. Weyburne and R. A. Brown, J. Cryst. Growth **77**, 136 (1986).

37. H. Moffat and K. F. Jensen, J. Cryst. Growth **77**, 108 (1986).

38. K. L. Hess, P. D. Dapkus, H. M. Manasevit, T. S. Low, B. J. Skromme and G. E. Stillman, J. Electron. Mat. **11**, 1115 (1982).

39. S. K. Shastry, S. Zemon and P. Norris, **Proceedings of the Ninth International Symposium on Gallium Arsenide and Related Compounds**, (Las Vegas, 1986), p. 81.

40. M. E. Givens, L. J. Mawst, C. A. Zmudzinski, M. A. Emanuel and J. J. Coleman, Appl. Phys. Lett. **50**, 301 (1987).

41. S. J. Jeng, C. M. Wayman, G. Costrini and J. J. Coleman, Mat. Lett. **2**, 359 (1984).

42. K. Takamura, J. Ogawa, K. Akimoto, Y. Mori and C. Kojima, Appl. Phys. Lett. **50**, 1149 (1987).

43. S. C. Palmateer, W. J. Schaff, A. Galuska, J. D. Berry and L. F. Eastman, Appl. Phys. Lett. **42**, 183 (1983).

44. American Society for Testing and Materials Annual Book of ASTM Standards, Part 3, (1979).

45. A. Mircea, R. Mellet, B. Rose, P. Daste and G. Schiavini, J. Cryst. Growth **77**, 340 (1986).

46. M. Razeghi, M. A. Poisson, J. P. Laravain and J. P. Duchemin, J. Elect. Mat. 12, 371 (1983).

47. Johnson Matthey, Inc. HP (8360) Series Hydrogen Purifier Operating Manual.

48. M. L. Cotton, N. D. Johnson and K. G. Wheeland, Met. Soc. Canadian Inst. Mining **Ann. Vol.**, 205 (1977).

49. K. L. Hess and R. J. Riccio, J. Cryst. Growth **77**, 95 (1986).

50. R. L. Messham and W. K. Tucker, J. Cryst. Growth **77**, 101 (1986).

51. H. M. Manasevit and W. I. Simpson, J. Electrochem. Soc. **116**, 1725 (1969).

52. C. F. Schaus, J. R. Shealy, L. F. Eastman, B. C. Cooman and C. B. Carter, J. Appl. Phys. **59**, 678 (1986).

53. R. S. Sillmon, N. Bottka, J. E. Butler and D. K. Gaskill, J. Cryst. Growth **77**, 73 (1986).

54. C. Van Opdorp and M. R. Leys, J. Cryst. Growth **84**, 271 (1987).

55. C. Blaauw and C. J. Miner, J. Cryst. Growth **84**, 191 (1987).

56. C. Murray, Semiconductor International (October 1985), pp. 72-78.

57. T. Nakagawa, H. Imamoto, T. Sakamoto, T. Kojima, K. Ohta, N. J. Kawai, Electron. Lett. **21**, 882 (1985).

58. H. S. Bennet, J. Appl. Phys. **60**, 2866 (1986).

59. L. J. Mawst, M. E. Givens, C. A. Zmudzinski, M. A. Emanuel and J. J. Coleman, IEEE J. Quantum. Electron. **QE-23**, 696 (1987).

60. S. M. Sze, **Physics of Semiconductor Devices**, (John Wiley and Sons, New York, 1981), pp. 637-678.

61. K. Hess, H. Morkoç, H. Shichijo and B. G. Streetman, Appl. Phys. Lett. **35**, 469 (1979).

62. A. A. Grinberg, A. Kastalsky and S. Luryi, IEEE Trans. Electron. Dev. **ED-34**, 409 (1987).

63. L. Esaki, IEEE Trans. Electon. Dev. **ED-23**, 644 (1976).

64. R. Tsu and L. Esaki, Appl. Phys. Lett. **22**, 562 (1973).

65. T. C. L. G. Sollner, E. R. Brown, W. D. Goodhue and H. Q. Lee, Appl. Phys. Lett. **50**, 332 (1987).

66. G. W. Taylor, R. S. Mand, J. G. Simmons and A. Y. Cho, Appl. Phys. Lett. **50**, 338 (1987).

67. T. K. Higman, J. M. Higman, M. A. Emanuel, K. Hess and J. J. Coleman, J. Appl. Phys. **62**, 1495 (1987).

68. M. A. Emanuel, T. K. Higman, J. M. Higman, J. M. Kolodzey, J. J. Coleman and K. Hess, Proceedings of the Fifth International Conference on Hot Carriers in Semiconductors, Boston, 1987.

69. T. K. Higman, M. A. Emanuel, J. J. Coleman, S. J. Jeng and C. M. Wayman, J. Appl. Phys. **60**, 677 (1986).

70. J. E. A. Whiteaway and C. Eng, IEE Proceedings **130**, 165 (1983).

71. A. R. Bonnefoi, D. H. Chow and T. C. McGill, J. Appl. Phys. **62**, 3836 (1987).

72. H. Kroemer, Appl. Phys. Lett. **46**, 504 (1985).

73. H. C. Casey and M. B. Panish, **Heterostructure Lasers, Part A**, (Academic Press, New York, 1978), p. 193.

74. I. Hase, H. Kawai, K. Kaneko and N. Watanabe, J. Appl. Phys. **59**, 3792 (1986).

75. K. Brennan, IEEE Trans. Electron. Devices **ED-32**, 2197 (1985).

76. W. Shockley, **Electrons and Holes in Semiconductors**, (D. van Nostrand Company, Inc., New York, 1950), pp. 258-264.

77. J. P. Andre, M. Boulow and A. Mircea-Roussel, J. Cryst. Growth **55**, 192 (1981).

78. M. J. Tsai, M. M. Tashima and R. L. Moon, J. Electron. Mater. **13**, 437 (1984).

79. T. F. Kuech and E. Veuhoff, J. Cryst. Growth **68**, 148 (1984).

80. J. M. Higman, private communication

81. S. D. Hersee, M. A. Diforte Poisson, M. Baldy and J. P. Duchemin, J. Cryst. Growth **55**, 53 (1981).

82. A. Kastalsky and S. Luryi, IEEE Electron Device Letters **EDL-4**, 334 (1983).

83. A. Kastalsky, R. Bhat, W. K. Chan and M. Koza, Solid State Electronics **29**, 1073 (1986).

84. A. M. Belyantsev, A. A. Ignatov, V. I. Piskarev, M. A. Sinitsyn, V. I. Shashkin, B. S. Yavich and M. L. Yakovlev, JETP Lett. **43**, 439 (1986).

85. O. A. Mezrin and S. I. Troshkov, Sov. Phys. Semicond. **20**, 819 (1986).

86. V. I. Tolstikhin, Sov. Phys. Semicond. **20**, 1375 (1986).

87. Zh. Alferov, O. A. Mezrin, M. A. Sinitsyn, S. I. Troshkov and B. S. Yavich, Sov. Phys. Semicond. **21**, 304 (1987).

88. A. M. Belyantsev, V. I. Gavrilenko, A. A. Ignatov, V. I. Piskarev, V. I. Shashkin and A. A. Andronov, Proceedings of the Fifth International Conference on Hot Carriers in Semiconductors, Boston, 1987.

89. J. Kolodzey, J. Laskar, T. K. Higman, M. A. Emanuel and J. J. Coleman, unpublished data.

Other Noyes Publications

SEMICONDUCTOR MATERIALS AND PROCESS TECHNOLOGY HANDBOOK

For Very Large Scale Integration (VLSI) and Ultra Large Scale Integration (ULSI)

Edited by

Gary E. McGuire

Microelectronics Center of North Carolina

This handbook is a broad review of semiconductor materials and process technology, with emphasis on very large scale integration (VLSI) and ultra large scale integration (ULSI). The technology of integrated circuit (IC) processing is expanding so rapidly that it can be difficult for the scientist working in one area to keep abreast of developments in other areas of the field. This handbook solves this problem by bringing together "snapshots" of the various aspects of the technology.

The generally accepted goals of VLSI are to produce devices with 100K gates, or memory bits, per circuit, or devices with geometries of less than 1 μm. The goals of VLSI technology are a natural evolution of current IC technology with the horizontal dimensions, gate length, junction depth and oxide thickness decreasing with each succeeding generation of product.

The methodology necessary for achieving the goals of VLSI is an extension of available lithographies used for pattern definition. However, new process technologies have been developed to minimize process-induced degradation of the defined geometry. In addition, new materials have been developed to meet the increasing demands that result from the decreasing geometries.

Many of these developments have never before been reviewed in a formal text, especially in one that discusses all major aspects of device processing. Several areas discussed in the handbook, such as orientation dependent etching, have remained trade secrets until just the last few years. There is also a section on packaging which is an area which has often been neglected but which is becoming an increasingly complex technology.

CONTENTS

ISBN 0-8155-1150-7 (1988) 675 pages